CLOSE READING THE ANTHROPOCENE

Reading poetry and prose, images and art, literary and critical theory, science and cultural studies, *Close Reading the Anthropocene* explores the question of meaning, its importance and immanent potential for loss, in the new geological epoch of the Anthropocene. Both close reading and scientific ecology prioritize slowing down and looking around to apprehend similarities and differences, to recognize and value interconnections. Here "close" suggests careful attention to both the reading subject and read "object." Moving between places, rocks, plants, animals, atmosphere, and eclipses, this interdisciplinary edited collection grounds the complex relations between text and world in the environmental humanities.

The volume's wide-ranging chapters are critical, often polemical, engagements with the idea of the Anthropocene and the changing conversation around reading, interpretation, and textuality. They exemplify a range of work from across the globe and will be of great interest to scholars and students of the environmental humanities, ecocriticism, and literary studies.

Helena Feder is Associate Professor of Literature and Environment at ECU, and the author of *Ecocriticism and the Idea of Culture* (2014/2016) and many articles, essays, interviews, and poems. She is the editor of several special journal issues and two books: *You Are the River* and this volume, *Close Reading the Anthropocene*.

"It's hard to imagine a better prescription for this fraught moment than: pay attention. The writers in this volume do just that, and thus help lay the groundwork for the only thing we need more than attention – action!"

– *Bill McKibben, author of* Falter: Has the Human Game Begun to Play Itself Out?

"*Close Reading the Anthropocene* offers a deeply intelligent introduction to an excruciatingly timely question: what place, if any, does reading have in a time of environmental catastrophe? An essential book for anyone interested in literature and environment studies."

– *Paul Outka, author of* Race and Nature from Transcendentalism to the Harlem Renaissance, *Associate Professor, The University of Kansas, US*

"Observed with the eyes of literature, the Age of the Human is something more than a geological hypothesis: it is a huge planetary novel, in which climate change, mass extinctions, yawning injustices, and global pandemics are interdependent themes all linked to the deep impact of our species. With *Close Reading the Anthropocene*, Helena Feder and her stellar ensemble of authors add to this story the many strata of literary imagination, providing both an indispensable archive and an inspiring blueprint for survival."

– *Serenella Iovino, Professor of Italian Studies and Environmental Humanities, University of North Carolina at Chapel Hill, Editor of* Material Ecocriticism and Environmental Humanities: Voices from the Anthropocene

CLOSE READING THE ANTHROPOCENE

Edited by Helena Feder

Routledge
Taylor & Francis Group
LONDON AND NEW YORK

First published 2021
by Routledge
2 Park Square, Milton Park, Abingdon, Oxon OX14 4RN

and by Routledge
605 Third Avenue, New York, NY 10158

Routledge is an imprint of the Taylor & Francis Group, an informa business

British Library Cataloguing-in-Publication Data
A catalogue record for this book is available from the British Library

Library of Congress Cataloging-in-Publication Data
Names: Feder, Helena, editor.
Title: Close reading the Anthropocene / edited by Helena Feder.
Description: New York, NY : Routledge, 2021. | Includes bibliographical references and index. Identifiers: LCCN 2021000203 (print) | LCCN 2021000204 (ebook) | ISBN 9780367466602 (hardback) | ISBN 9780367466596 (paperback) | ISBN 9781003030270 (ebook)
Subjects: LCSH: Nature in literature. | Human ecology in literature. | Ecocriticism. | Nature—Effect of human beings on. | Human ecology and the humanities. | Environment (Aesthetics)
Classification: LCC PN48 .C55 2021 (print) | LCC PN48 (ebook) | DDC 809/.9336—dc23
LC record available at https://lccn.loc.gov/2021000203
LC ebook record available at https://lccn.loc.gov/2021000204

ISBN: 978-0-367-46660-2 (hbk)
ISBN: 978-0-367-46659-6 (pbk)
ISBN: 978-1-003-03027-0 (ebk)

Typeset in Bembo
by Apex CoVantage, LLC

CONTENTS

List of contributors *vii*
List of figures *x*
Acknowledgments *xi*

Introduction: the unbearable closeness of reading 1
Helena Feder

1 Inhabiting words, inhabiting worlds: a case for pragmatist close reading 15
Amelia Marini

2 Ecopoetics and the myth of motivated form 30
Greg Garrard and Rina Garcia Chua

3 Assembling the archive: close(ly) reading great auk extinction with Walton Ford 45
Nicole M. Merola

4 Bartleby and the politics of measurement 60
Helena Feder

5 Close reading at the end of time 76
Mark C. Long

6 Postcolonial anthropocene and narrative archeology in *Burma Boy* 93
Senayon Olaoluwa

7 Key west in the anthropocene: Stevens and Bishop close reading Florida 102
Peter Balaam

8 The tree as archive: George Nakashima and the nuclear age 116
Isabel Duarte-Gray

9 Going underground: in defense of deep reading 131
Graham Huggan

10 Reading in the dark: the aura of eclipse 144
Hilary Thompson

11 John Masefield's "The Passing Strange": derangements of scale 159
Timothy Clark

12 From scale to antagonism: reading the human in Kurt Vonnegut's *Galápagos* 175
C. Parker Krieg

Index *190*

CONTRIBUTORS

Peter Balaam teaches American Studies and English at Carleton College in Northfield, Minnesota. He is the author of *Misery's Mathematics* (Routledge 2009) and is at work on a study of form and limits in the thought of Ralph Waldo Emerson.

Timothy Clark is Professor emeritus of English at the University of Durham, UK, and a specialist in the fields of modern literary theory, continental philosophy (especially the work of Martin Heidegger and Jacques Derrida), and environmental criticism. His most recent books are *Ecocriticism on the Edge: The Anthropocene as a Threshold Concept* (Bloomsbury, 2015) and *The Value of Ecocriticism* (Cambridge University Press, 2019). He has edited special issues of *The Oxford Literary Review* on "Deconstruction, Environmentalism, and Climate Change" (June 2010), "Deconstruction and the Anthropocene" (December 2012), "Overpopulation" (June 2016), and, most recently (with Jennifer Ford), "Deconstruction and the Child: Children's Literature: Anthropomorphism: Animality: Posthumanism" (December 2019).

Isabel Duarte-Gray is a Ph.D. candidate in the English Department at Harvard University, where she specializes in Latinx literature and twentieth-century American Ethnic Imaginaries. Her work has appeared or is forthcoming in collections on the Anthropocene and Southern Studies. Her first collection of poetry debuts with Sarabande Books in May 2021.

Helena Feder is Associate Professor of Literature and Environment at ECU, and the author of *Ecocriticism and the Idea of Culture* (2014/2016) and many articles, essays, interviews, and poems. She is the editor of several special journal issues and two books: *You Are the River* (NCMA 2021) and this volume, *Close Reading the Anthropocene* (Routledge 2021).

Rina Garcia Chua is currently a Ph.D. candidate in Interdisciplinary Studies in the University of British Columbia Okanagan. She is the editor of *Sustaining the Archipelago: An Anthology of Philippine Ecopoetry* (2018) and is working on an edited collection, *Empire and Environment: Confronting Ecological Ruination in the Transpacific*, with Heidi Hong, Jeffrey Santa Ana, and Xiaojing Zhou, forthcoming with the University of Michigan Press in 2021. She is also the Diversity Co-Officer for the Association for the Study of Literature and the Environment (ASLE) and Poetry Editor of *The Goose: A Journal of Arts, Environment, and Culture in Canada.*

Greg Garrard is Professor of Environmental Humanities and Associate Dean of Research in the Faculty of Creative and Critical Studies, UBC Okanagan. He is the author of *Ecocriticism* (Routledge 2004, 2011, 2nd ed), essays on eco-pedagogy, animal studies, and environmental criticism, and coauthor of *Climate Change Skepticism: A Transnational Ecocritical Analysis* (Bloomsbury Academic 2019). He is the editor of *The Oxford Handbook of Ecocriticism* (OUP 2014) and series coeditor of *Environmental Cultures* from Bloomsbury Academic.

Graham Huggan teaches in the School of English at the University of Leeds (UK). His research straddles three fields: postcolonial studies, tourism studies, and environmental humanities. Published work that similarly combines these fields includes *Postcolonial Ecocriticism* (Routledge/Earthscan, 2010), coauthored with Helen Tiffin, and *Colonialism, Culture, Whales* (Bloomsbury, 2018). He is currently working on a coauthored book on modern British nature writing and on an international research project on European transboundary national parks.

C. Parker Krieg teaches in the Global Studies program at the University of Nebraska Lincoln's School of Global Integrative Studies. He previously held a postdoctoral fellowship in environmental humanities at the University of Helsinki, affiliated with the Faculty of Arts and the Helsinki Institute of Sustainability Science. His research and teaching focuses on post-Fordism and the environment in American literature, especially as it relates to cultural memory and environmental justice.

Mark C. Long is Professor of English at Keene State College, where he teaches courses in American literature and culture, nonfiction writing, poetry and poetics, and the environmental humanities. His most recent work includes the co-edited collection of essays, *Approaches to Teaching the Works of Ralph Waldo Emerson*, published by the Modern Language Association. Mark is associate editor for the journal *Pedagogy: Critical Approaches to Teaching Literature, Language, Composition, and Culture.*

Amelia Marini is Assistant Professor at Cuesta College in San Luis Obispo, California. A generalist with a focus on American literature, she works with texts spanning from the seventeenth to twenty-first centuries, exploring the intersections of literary friendship, American pragmatist philosophy, and the phenomenology of reading.

Nicole M. Merola is Professor of environmental humanities and American literatures in the Department of Literary Arts and Studies at the Rhode Island School of Design. Her teaching and research interests include Anthropocene studies, biodiversity and extinction studies, climate change cultures, environmental affect, and theories of nature culture.

Senayon Olaoluwa is Acting Director of the Institute of African Studies, University of Ibadan, Nigeria. His article "Dislocating Anthropocene: The City and Oil in Helon Habila's *Oil on Water*" was recently named among the top ten on environmental justice by Oxford University Press and *ISLE: Interdisciplinary Studies in Literature and Environment.*

Hilary Thompson is Associate Professor of English at Bowdoin College, Maine. Her research and teaching bring together the fields of nonhuman animal studies and contemporary global anglophone fiction. She is the author of *Novel Creatures: Animal Life and the New Millennium* (Routledge 2018). Her current project *Worldly Spirits: Extra-Human Dimensions and the Global Anglophone Novel* looks at animal spirits and extra-human dimensions in recent fiction.

FIGURES

8.1	Butterfly keys, clearly visible as George Nakashima inspects the Altar for Peace at the Cathedral of St. John the Divine.	121
8.2	Nakashima displays the Altar for Peace in his workshop in New Hope, Pennsylvania.	125
8.3	An overhead view of the first Altar for Peace at the Cathedral of St. John the Divine.	126
10.1	*The Arts of Living on a Damaged Planet* conference website.	152

ACKNOWLEDGMENTS

I would like to thank the contributors to this volume, several of whom were involved in the project from its earliest days as an Association for the Study of Literature and the Environment panel at the American Literature Association Annual Conference, and my colleagues John Parham and Sarah McFarland for allowing me to quote from their forthcoming books. I am very grateful to ASLE and the ALA, my editors at Routledge, and all those working in ecocriticism and the environmental humanities. I would also like to thank Walton Ford, for kindly granting permission to use "Funk Island" for the cover of this book, and Nancy Pagh, for granting permission to quote from her poem, "Moon Jelly," in Chapter Two; every effort was made to contact the estate of John Smith for permission to quote from "The Birds Returned" in the same chapter. To my friend and mentor, the writer and photographer David Robertson, and to my family, I am ever grateful for their love and support. I am particularly indebted to my son, Hugh Feder-Clarke, for whom I hope to leave this world better than I found it.

INTRODUCTION

The unbearable closeness of reading

Helena Feder

Particularly, we find, in the Anthropocene.

As Milan Kundera asks in Chapter 2 of *The Unbearable Lightness of Being*, "If eternal return is the heaviest of burdens, then our lives can stand out against it in all their splendid lightness. But is heaviness truly deplorable and lightness splendid?" For, you see,

> The heavier the burden, the closer our lives come to the earth, the more real and truthful they become. Conversely, the absolute absence of a burden causes a man to be lighter than air, to soar into the heights, to take leave of the earth and his earthly being, and become only half real, his movements as free as they are insignificant. What then shall we choose? Weight or lightness?
>
> *(5)*

Italo Calvino's *Six Memos for the Next Millennium*, written in 1985, a year after the publication of Kundera's novel (and not long after Eugene F. Stoermer began using the term Anthropocene),[1] is a meditation on the qualities of literature that begins with a chapter upholding "the values of lightness" (3). In his preface to the book, Calvino notes that as the end of the millennium approaches, people have begun to "wonder what will happen to literature and books in the so-called postindustrial era of technology. . . . My confidence in the future of literature consists in the knowledge that there are things that only literature can give us, by means specific to it" (i).

A page before the passage I quoted from Kundera's novel, he writes, "In the sunset of dissolution, everything is illuminated by the aura of nostalgia, even the guillotine" (4). This aura is an effect of what seems the impossibility of judgment in a world that is "ephemeral, in transit" (4). Does this lightness of perpetual loss conflate nostalgia and meaning, make nostalgia – a form opposed to critical thought – the only meaning available to us? Moving further in reverse, back to the very first page, Kundera articulates Nietzsche's idea of eternal return negatively, explaining that "life which does not return is like a shadow, without weight, dead in advance, and whether it was horrible, beautiful, or sublime, its horror, sublimity, and beauty mean nothing," with the example: "We need take no more note of it than a war between two African kingdoms in the fourteenth century, a war that altered nothing in the destiny of the world, even if a hundred thousand blacks perished in excruciating torment" (3). This of course begs the question: how could any war alter "nothing in the destiny of the world?" This passing illustration of the paradox of Nietzsche's terrible "lightness" calls into question both the "we" of its sentence and its conception of "the world." Any war must matter, and perhaps a very great deal, to the descendants of those who fought in it – to those who won and those who lost and, if the victors enslaved the vanquished, to any who benefited or lost from the capitalization of human beings. It would be of material significance, whether generations down the line realize it or not. Kundera's close reading, his writing of this idea of eternal return, seems to misread its own example. Is it standing too close or too far away to realize its own position in relation to that of others, including its readers? Or has Kundera's immersive novel staged this antagonism, cultivating a visceral *and* critical tension between text, subject, and subtext?

Even a brief close reading of these passages opens two of the key contentions surrounding the idea of the Anthropocene: the question of meaning and the "we" of the Anthropos. Meaning is not separate from the matter of who, human or nonhuman, or indeed what, signifies; it is not separate from the humanist ideology of culture. In the Age of "Man," the question of signification may all too easily reify the systems it claims to challenge, if not describe. "We" do not bear the burden of responsibility or suffer the consequences of ecological catastrophe equally, as the name of this geological era would suggest. As Gary Snyder has said, the term blames the entire species for the errors of white male capitalists.[2] And, as Kathryn Yusoff has it in *A Billion Black Anthropocenes or None*, "imperialism and ongoing (settler) colonialism have been ending worlds for as long as they have been in existence"; the Anthropocene "is just now noticing the extinction it has chosen to continually overlook in the making of its modernity and freedom" (xiii).

While the term elides the social stratification and forms of exploitation that created and continues to create the circumstances it articulates, the Anthropocene is also a means of giving conceptual reach and weight to the ecological damage caused by human activity, by imperial expansion and capitalist "development." In May 2019, the Anthropocene Working Group,[3] voted to treat the Anthropocene "as a formal chrono-stratigraphic unit defined by a GSSP," a Global boundary Stratotype Section and Point, "colloquially known as a 'golden spike,'" with "one of the stratigraphic signals around the mid-twentieth century of the Common Era" as the primary guide (n.p.).[4] The middle of the Twentieth Century coincides with an array of signals "resulting from the 'Great Acceleration' of population growth, industrialization and globalization" (n.p.).

While this was a compelling temporal marker for the AWG (with 88% voting in favor of this position), dating the onset of the Anthropocene has been almost as controversial as the term itself. A few years before the AWG vote, Simon Lewis and Mark Maslin of the University College London argued the case for 1610:

> The year 1610 marks the low-point of a dip in global carbon dioxide levels caused by a drastic reduction in farming in the Americas. This was a knock-on effect of the 50 million or so indigenous deaths that resulted from the introduction of smallpox to the continent by European colonialists. A secondary marker of the colliding of worlds is the sudden appearance, in 1600, of fossil pollen maize, a Latin American species, in the European marine record.
>
> *(n.p.)*[5]

There remain other views, such as that of William Ruddiman, who argues that the Anthropocene is "a gradual change, a slow transformation of the planet that began 5,000 years ago . . . driven primarily by agriculture. Before the year 1750 . . . humans had already cleared so much forest as to produce 300 billion tons of carbon emissions. Since 1950, deforestation has only led to 75 billion tons of emissions" (n.p.).[6] From the geological strata to their social and ecological implications, the Anthropocene is, itself, the subject and object of close reading.

Whether you're sticking with the Holocene, or find Plantationocene,[7] Homogenocene,[8] or Capitalocene[9] more useful terms, the Anthropocene, like the damage it purports to describe, isn't going away anytime soon – if ever. The term has reached almost every discipline in the sciences and humanities, every scientific institution, and the popular imagination. It's part of the ecosphere, the unbearable lightness of too much carbon dioxide, too much information and misinformation, and too little agency.

If we think and act to give our lives weight, to give them meaning, one of the ways we do so is with the lightness of literature. I write lightness after Calvino's fashion, in which literature's characteristics also contain or give way to their opposites: lightness and weight, quickness and slowness, exactitude and mystery, visibility and invisibility, multiplicity and unity, and – Calvino's unwritten chapter – consistency and variation. Close reading is a process of not only parsing but also recognizing the connections and complexities between these seeming opposites, including those that also apply to scholarship, such as form and content, content and context, coherence and contradiction, etc.

There is no need to recount the critical history of close reading in this introduction (though many chapters in this volume insightfully address the subject). This work has been done well by others, perhaps most recently by Frank Lentricchia and Andrew DuBois in their anthology *Close Reading*, which chronicles the term and practice from John Crowe Ransom and Cleanth Brooks to Eve Sedgwick and Homi Bhabha. DuBois's introduction is particularly good at contextualizing the complex politics of no less a subject than "the issue of reading in modern literary criticism" (2). He demonstrates that attending to a text was never one universal practice or a practice of decoding universal truths. Following Barthes, he writes that "universality is simultaneity of value," which, if it ever existed, "would risk making interpretive reading impossible, and would render the reader superfluous . . . the critic, the reader of distinction, may resist this situation not only from political commitment, but also from a readerly and writerly commitment" (23).

While close reading is, as we know, always political and historical even when it is expressly apolitical or ahistorical, close reading itself is not always or only "close." One form of distant reading repositions literature, art, popular and material culture within networked histories of global human and nonhuman events, scaling up and out, or down and in, to see unexpected lines of connection between text and layers, sediments, of context. And while Franco Moretti has suggested his version of "distant reading," the quantitative analyses of literature and literary forms with algorithms (imagine haystacks of needles), should supplant close reading,[10] it was in a sense already – however alien it may seem to its parent – part of the varied, multifarious ecology of close reading.

And here we come to a convergence of close reading and scientific ecology – both prioritize slowing down and looking around to apprehend similarities and differences, to recognize and value interconnections. Here "close" suggests careful attention to *both* the reading subject and read "object." If attention is a form of love, then ecology is, as Snyder has written, "a problem of love" (4). And while the Sixth Mass Extinction isn't a problem that may be solved by close reading, I can't imagine it is one that will be meaningfully addressed without it.

The narrower question of the value of literary works as objects of sustained critical attention in the midst of overlapping ecological and social crises has, however, troubled some in the fields of ecocriticism and the environmental humanities. It is the newest, and perhaps the most convincing, version of the charge that "poetry makes nothing happen" ("In Memory of W. B. Yeats," section 2, line 5). As a reader and writer, I stand with Auden's reply to himself in line 10. Here, for context, is the whole section:

> You were silly like us; your gift survived it all;
> The parish of rich women, physical decay,
> Yourself: mad Ireland hurt you into poetry.
> Now Ireland has her madness and her weather still,
> For poetry makes nothing happen: it survives
> In the valley of its making where executives
> Would never want to tamper, flows on south
> From ranches of isolation and the busy griefs,

Raw towns that we believe and die in; it survives,
A way of happening, a mouth.

This poetry, which "makes nothing happen," is born from the pain of "mad Ireland," from the individual amid social struggles. The suffering continues but, here, poetry is a river that resists it. Flowing from the valley of embodied, embedded consciousness – from what isn't managed and legislated – it gathers isolated lives and collective griefs, raw antagonisms and the ideologies that sustain them, becoming – doubly – a *mouth*: that which speaks for individuals *and* the point at which this river enters an ocean. In becoming part of an ocean, part of a social and ecological body, poetry becomes part of a body *politic*. In this way, poetry is a "happening" in which social feeling is created and sustained, where individual and common humanity, unique and shared differences, may be articulated, felt, and understood. It is a means by which we may speak beautiful and difficult things to each other and, yes, truths to power.

While not all the authors in this volume close read literature or argue explicitly for the unique value of this enterprise, many demonstrate what remains to be learned and grown from the practice. As John Parham notes in his Introduction to the *Cambridge Companion to Literature and the Anthropocene,* "If the Anthropocene has engendered an inquisition into the value of literature and literary criticism it also offers an opportunity to reinvigorate both" (12). He argues that literature "could help effect a cumulative movement in *anthropos* towards an understanding that humans are embedded on the Earth . . . by adapting and evolving its practice in two ways: the sharing of divergent experiences (for different people, even species) of the Anthropocene; and by reconnecting human life with exponentially vaster scales: deep history, the planet Earth, the distant future" (12–13).

Aside from its uses in the other disciplines, close reading sits at the intersection of three literary activities: creative, critical, and pedagogical. While the last two are part of an obvious continuum, the former may be forgotten by those of us who predominantly write secondary and teach primary texts. The act of writing poetry, fiction, and nonfiction is itself a process of close reading, just as reading literature is an integral part of the process by which its meanings are made. Writing is close reading the world – articulating and delineating relations, exploring and developing connections.

But in the Anthropocene, this moment often called unthinkable, the world itself feels *unbearable*. In the same month the AWG voted on the Anthropocene in 2019, the UN Intergovernmental Science-Policy Platform on Biodiversity and Ecosystem Services reported that our climate has changed enough that "grave impacts on people around the world [are] now likely" (n.p.).[11] Of course, this age of human exceptionalism also endangers a staggering number of other animals and plants.[12] As Elizabeth Kolbert wrote 5 years before this report, our accelerating rate of multispecies extinction is unprecedented in human history, and "predicted to be the most devastating extinction event since the asteroid impact that wiped out the dinosaurs" (n.p.).

Reading in these times, whether it is the news or a new novel, may feel unbearable not only because we're facing apocalypse, but also because we're faced with our own resistance to human extinction. As Sarah E. McFarland writes in *Ecocollapse Fiction and Cultures of Human Extinction*, "climate change fiction as a genre of apocalyptic and post-apocalyptic writing has resisted facing the potentiality of human species extinction, following instead traditional generic conventions that display a predilection for happy endings by imagining primitivist communities of human survivors with the means of escaping the consequences of global climate change" (2). Close reading, then, becomes more necessary, as "forecasting human extinction necessitates coming to terms with the complex mapping of ecosystems, global warming, and human and nonhuman cultures to exert pressure on human exceptionalism's contradictions, exclusions, repressions, and marginalizations" (10). The more distant we wish to be from what Robinson Jeffers called "the monster's feet,"[13] the closer we should look at our constructions of the human in the context of the more-than-human world.

Chapters

Reading poetry and prose, images and art, literary and critical theory, the essays in this volume attempt to do just this, to explore the unbearable being of meaning, its importance and immanent potential for loss, in the Anthropocene. Moving between places, rocks, plants, animals, atmosphere, and eclipses, these close readings are grounded in the complex relations between text and world. They are critical, and sometimes polemical, engagements with the question of the Anthropocene and the changing conversation around reading, interpretation, and textuality. Individual, idiosyncratic, collaborative, or collective, they exemplify a range of work in ecocriticism and the environmental humanities from across the globe (North America, Africa, and Europe). As you will see, many chapters have shared concerns, and several situate their readings in a shared constellation of critical and theoretical works.

Amelia Marini's "Inhabiting Words, Inhabiting Worlds: A Case for Pragmatist Close Reading" is based on the idea that "climate wisdom might indeed be a matter of learning to inhabit words and the worlds we compose with them. Within the Anthropocene, we've forgotten that songs and stories, in addition to eulogizing, have long been used to call forms of life into being." Developing a notion of "'pragmatist" close reading inspired by the philosophy of William James, Marini reminds us that "seeing and saying are looped processes: we do not only say what we see, we also tend to see the concepts we've learned to say." Personal and scholarly, this chapter challenges us to "begin to conceive of climate change denial as primarily a problem for the humanities," calling for "a poesis of climate change: a way to think about how the reader or receiver of 'information' *makes* something from it." Close reading, as Marini has it, may teach us "to better understand the connection between seeing and saying," presenting "an opportunity to inhabit texts in individual and personally resonant ways that require intersubjective imagining and relational thinking: the very skills that we must learn in order to live well upon this planet."

Jumping into the fray of old and new debates about literary interpretation and extraliterary meaning, Greg Garrard and Rina Garcia Chua's essay, "Ecopoetics and The Myth of Motivated Form," enacts "a principled curiosity" about those who mistakenly believe that "poetic forms have meaning and purpose *independent of their semantic content.* We will be calling this error 'the myth of motivated form.'" Wryly, they observe that "claims about the motivation of literary form repeat like Groundhog Day. 'Free verse,' or open form poetry, has been aligned repeatedly with various axes of human emancipation, while the changing form of the novel inspired decades of debate through the middle of the 20th century about the ideological valence of realist and modernist alternatives." Ecopoetry, they argue, has been the most recent victim of such "instrumentalizing readings and writings" for the political needs of the moment. Moving from I.A. Richards to recent ecopoetic theory, this chapter contrasts strong and modest claims about the significance of form, testing them in relation to Canadian poet John Smith's "The Birds Returned" and American poet Nancy Pagh's "Moon Jelly" to consider the implications for the practice of close reading.

Beginning by surveying the "flourishing" of extinction studies in the environmental humanities, Nicole Merola's "Assembling the Archive: Close(ly) Reading Great Auk Extinction with Walton Ford" argues that his "extinction-focused objects rehearse located and historicized Anthropocene stories; to use Elizabeth DeLoughrey's formulation, he provincializes the Anthropocene." In this sense, Ford's work, featured on the cover of this volume, not only brings the Anthropocene "closer" but serves "as examples for modes of eco-studies reading in/of/for the Anthropocene." The "close(ly)" in Merola's title signals the value she places on "extended attention" to Ford's paintings and her "investment in ongoing debates among literary critics about 'proper' reading methods for approaching our objects: close, contextualized, descriptive, distant, formalist/new formalist, historicized, ideological, reparative, surface, symptomatic." She argues that by bringing particular moments in the extinction of the great auk into view, Ford presents a new way to periodize the Anthropocene; he expands the kind of stories told with this rubric by "offering dates tied to the exploitation of nonhuman animal bodies, rather than dates tied to human inventions, for instance the steam engine or atomic bomb, which regularly figure as potential golden spikes." Merola's important, timely move to recontextualize the Anthropocene in the nonhuman is echoed in Chapter 8, which also close reads visual art.

Like many others, my own chapter in this volume, "Bartleby and the Politics of Measurement," asks what it means to read closely, to discern meaning in an age at once characterized by a new awareness of vertiginous scale and complex interconnection and by the collapse of multispecies agency into a singularity, an idea of "Man." The world is, of course, more-than-human and more than our measure of it. Scale itself expands and recedes almost infinitely, while "close" has become uncomfortably closer – from toxins inherited in utero[14] to cultivated misinformation sweeping the globe faster than any disease[15] – and the human future now often seems behind us. Moving from the discomfort of scale to that of proximity, this chapter treats close reading as a critical and creative practice, demonstrating the way in which the reality of climate change may change the way we think about even the most persistent, indeed sedimentary, works of American literature. Reading Melville's "Bartleby, the Scrivener" in these times in which "Man" seems "to become his own ghost, an uninvited guest," may "provoke a consideration of the ethics of inhabitation in human and inhuman, co-evolutionary and co-constitutive, terms." On and beyond Wall Street, "Bartleby," I argue, "dramatizes the bad faith of humanism as the bad faith of capitalism and, in the context of close reading the Anthropocene, invites speculation about agency, islands, and becoming rock."

Mark Long's "Close Reading at the End of Time" proposes that "the reading we do beyond the parochial concerns of scholarly communities, that we enact in proximity to others, may be the closeness of reading that matters most." Weaving together an impressive array of critical and theoretical sources, his chapter surveys and considers recent debates about reading in "studies concerned with past and present ways of thinking about nature. Of particular interest will be critical projects preoccupied with scales of time . . . or what I am calling 'close reading at the end of time.'" Redefined as a relational activity, close reading offers "a collaborative response" to ecological crisis, the hope that those "cultivated in the practice of close reading together would engage more people with the cultural problem of attention to the particular urgencies of living in our fragile and precarious world." In this way, Long argues that the practice presents "an ethical response to the challenge of the Anthropocene," creating in readers "ways of thinking and habits of mind to orient themselves toward one another, to the past, and to the planetary future we share."

Demonstrating that postcolonial studies has always been a matter of unearthing and contextualizing what remains, Senayon Olaoluwa's "Postcolonial Anthropocene and Narrative Archaeology in *Burma Boy*" challenges universalizing constructions of the Anthropocene with the persistence of material bodies – child soldiers – extracted and exploited like any other "natural resource" for use by western capitalism. Situating his close reading of Biyi Bandele's novel between Kathryn Yusoff's "White Geology" and Richard Young's "postcolonial remains," Olaoluwa argues that coming to terms with postcolonial fiction "in the broader context of the Anthropocene must begin with paying attention to the specific history it evokes." This close reading is, he implies, key for transformational consciousness, reminding us that the work of geology "operates not only through extraction but also through conscious and unconscious covering, the submergence of the extractive remains at the site of exploitation. This is what I term *the afterlife of geology*." As a counter to this extractive logic, narrative archeology discovers and rediscovers the "remains of colonial violence with its numerous resonances for the present and the future in the postcolony." Olaoluwa's close reading of Bandele's novel excavates not only the "improper colonial burial of the past, but also the present humanist fiction buried in a universalizing notion of the Anthropocene."

Peter Balaam's "Key West in the Anthropocene: Stevens and Bishop Close Reading Florida" begins with the questions: How do we read poetic traditions rooted in the Holocene? And what have these poets to teach us about "the encounter of mind and nature in this era that *Anthropos* has made in its own image?" Grounding his lyrical exploration in the materiality of a specific ecosystem, Balaam argues that the Florida poems of Wallace Stevens and Elizabeth Bishop offer "a compelling portrait of modern Anthropos confronting the scene of diminished nature and close reading it. They serve as able guides to the disciplines of attention, descent, and losing that ought to typify that encounter." Resting on the promise, the possibility of such close attention, the central argument of this chapter complements those of Marini and Long: "To think realistically – modestly – about the value of close reading in the Anthropocene might begin by asking whether mere readerly attention, a stance or orientation that lies at the heart of close reading, might resonate with paradigms of ecological thinking. As a practice close reading was less about finding instances of irony, ambiguity, and tension in poems than it was a way of heeding art's invitation simply to notice." Readerly and writerly concerns are entangled with each other *and* the world, a world which has been terribly damaged, which "Lose(s) something everyday," but – if we look closely we'll see – is *not yet* lost to us.

In "The Tree as Archive: George Nakashima and the Nuclear Age," Isabel Duarte-Gray combines ecocriticism, critical plant studies, and Asian American studies to consider Nakashima's woodwork and writing as a praxis that may be close read as a rejection of anthropocentrism. Looking at archival correspondence between Nakashima and biologist George Wald, Duarte-Gray argues that he viewed his "Altars for Peace" not as static monuments but as living interventions for disarmament in the nuclear age – the very "golden spike" some argue for dating of the Anthropocene. She writes that in the shadow of Rey Chow's "age of the world target," Nakashima designed the Peace Altars to bring civilians and government representatives to the same table, employing "the medium of wood grain to place the natural world at the center of the nuclear discussion, to ask how civic action functions in an era of unprecedented, and even apocalyptic, global interdependence of individuals, nations, and living things." Close reading material as well as written texts, this chapter demonstrates how Nakashima saw, and made others see, trees as legible, as historical, and so resonant with meaning. Reading Nakashima's art, his own close reading of the world, as an ecocritical intervention suggests how the practice of close reading itself might broaden in the Anthropocene and, at the same time, recontextualize the human in the more-than-human world.

Graham Huggan's "Going Underground: In Defense of Deep Reading" makes a "firm push against those scholars, Bruno Latour prominent among them, who claim either that critique has become all too predictable, ossified in its operational manoeuvres, or that it has simply had its day." He argues that critique is "as likely to pick fault with itself as with the text, which is one of the many reasons it is valuable, while the received idea that critique is basically extractive . . . seems as over-schematic as the equal-and-opposite view that it is unconscionably naïve (or politically quietist) to look at patterns of linguistic diversity or complexity." Instead, critique is needed more than ever "in the emergent (also emergency) conditions of the Anthropocene, which affects all of us but unevenly." Much like Garrard and Chua's, Merola's, and Long's chapters, Huggan situates his reading in debates around close versus distant, symptomatic versus surface. In Robert Macfarlane's *Underland* and Will Hunt's *Underground*, both published in 2019, "an interplay between 'surface' and 'deep' levels is literalized," requiring methods of close reading that "mesh with the radical epistemologies that [Rosi] Braidotti references as a basis for confronting the Anthropocene predicament – epistemologies that reposition deep reading as a multilayered critique of modern technologies of extractivism rather than as a prescriptive form of extracting buried meaning from the text." Close reading these nonfiction works of subterranean exploration and imagination demonstrates the uncanny way in which they "mirror" each other and remind us of the unfinished history of exploitation that is an index of the Anthropocene. Like Olaoluwa, Huggan draws on Kathryn Yusoff's work (among others), to consider the "cloven," if not fractured, nature of geology and what it means to read, and live on, the ground.

In "Reading in the Dark: the Aura of Eclipse," Hilary Thompson offers a critically and theoretically luminous reading of literary "readings of heavenly bodies, particularly the dramatic instance in which a specific alignment briefly swallows us in darkness, reveal[ing] the paradoxes built into our ambivalent desires for both closeness and distance in interpretation." Both the idea of Anthropocene and the full solar eclipse offer a spectral sense "of planetary life's extinction; in the context of the former, the latter becomes more significant, as the basis for spectacular creativity, for a concentration of meaning in which the very terms of literariness come to light." Sensitive to the complexity of such flashes of light, Thompson proposes a practice of conjoined reading "to view these strands together, with an eye for the ways that the total solar eclipse suggests a glimpse of our destruction but also a chance to learn ways of seeing ourselves in darkness." This chapter close reads Annie Dillard's "Total Eclipse" and Anne Carson's "Totality: The Colour of Eclipse" – and, behind these coronas, Virginia Woolf's 1928 essay, "The Sun and the Fish" – in the context of Anna Tsing's hidden commons. Thompson's lyrical, self-reflexive practice inhabits a layered circuit, a world, in which to "scan for strange sparks of connectivity is to recognize that one reads in the dark."

Beginning with the assertion that "any clear distinction between what is 'close' and what is 'distant' has severely eroded," Timothy Clark's "'The Passing

Strange': Derangements of Scale" close reads the poem by John Masefield of the same title in relation to "scale critique." The chapter exemplifies how the sense of a new epoch may change how we read, making Masefield's poem "of renewed interest" for its content and form; "'The Passing Strange,' if immediately foreign to the 'Anthropocene' debate, reinforces arguments that it is in the resources of poetry, with its demands of close reading, that some of the scalar challenges of representation can be most effectively met." Clark argues that such reading is particularly useful "as a mode of attention that resists premature synthesis, which recognises ethical irresolution in refusing the complacency of the good conscience." Rather than arguing that Masefield's work anticipates the idea of the Anthropocene, Clark sees in "The Passing Strange" the revelation "of something that has always been in operation, of the human as part of a biological and ecological dynamic in which destruction and extinction, including self-extinction, have always been inherent, even constitutive and elemental."

Recognizing that the Anthropocene is "structured by an antagonism between capitalism and the earth, which is reproduced through dispossession, exploitation, and exclusion," C. Parker Krieg begins his chapter with a call for reading practices that "rescue the human from the double erasure of frameworks that elide social difference under the figure of the *anthropos* and, at the same time, bury the *anthropos* beneath geological, evolutionary, and atmospheric forces." "From Scale to Antagonism: Reading the Human in Kurt Vonnegut's *Galapagos*" reads Vonnegut's novel in the context of the author's "criticism of war, capitalism, and the complicity of professional class intellectuals in planetary crises" to consider its "satirical evolutionary narrative in relation to recent moves 'beyond' the human, either to digital disembodiment, or to post-historical animality." Alive to what I would call the politics of measurement, Krieg argues that close reading "must not reproduce sublime derangements of scale that symptomatically reinforce the double erasure, but should connect literature to social antagonisms that reproduce the Anthropocene as a common condition." This critique is rooted in the human, not as other-than-animal, but as the cause and site of suffering and the potential for its change. Nailing his colors to the mast, he writes if Fredric Jameson's much noted observation that it is "easier to imagine the end of planetary life than the end of capitalism is correct, then, surely, interpretive practices that erase the human are an accelerated route to this same impasse."

Close and distant

The many threads that run through these chapters, including those that present opposing arguments, suggest that close reading is still key to nuanced thought, and so thoughtful action – creative, critical, pedagogical, and ecological. Close reading the idea of the human is both a necessary and messy practice, particularly when the term "Anthropocene" seems to risk evacuating its politics through repetition, proliferation, and variety. Depending on its use, its turn in a text, it may signal a resistance to false holism, an awareness of our stratified humanity and damaged planetary ecology, or the shiny, technical face of domination.

The Covid-19 pandemic has, inevitably, inflected how many of us think of the words "close" and "distant," and – for some – made the politics of the present more present. Living in increasingly polarized countries, amid torrents of fascist rhetoric, partisan acrimony, and misinformation, in the ever-widening trench between the rich and the poor, we were already socially *distant*.

In this context, close reading is more than just another tool in the kit of critical practice. It is the insistence that texts are always in context, that words, images, and objects have complex, entangled meanings and histories which matter, that meaning itself is always embodied, even when printed on a page or digitized on a screen. While people live more unequally, indeed farther apart, than ever, we are also growing *unbearably* closer. Global ecological depredation, degeneration, and disaster remind us we are all in the same shrinking boat.

In the fifth chapter of *Six Memos for the Next Millennium*, "Multiplicity," Calvino articulates an idea of networks as an ecology (via Carlo Emilio Gadda's "system of systems," [105–06]), but one which may become too dense and too expansive – in which thought and text may multiply infinitely from their own "organic vitality" (110). The same may be said of what we call information and misinformation; "[k]nowledge as multiplicity" has shaped the modern and postmodern world (116). If Calvino had lived to write his last lecture, the missing sixth chapter on "Consistency" would, I believe, have also emphasized the importance of variety and variation: the unexpected leaps in the evolution of organisms, ideas, and cultures. Even in its absence, it is a reminder that despair and determinism only *seem* to define the future. From the heaviness that Calvino felt looking forward back in 1985, he invites us to rise to this challenge of "lightness": "Only if poets and writers set themselves tasks that no one else dares imagine will literature continue to have a function" (112). Yes, and readers too.

Notes

1 Described by Andrew Revkin in "Confronting the 'Anthropocene" in *The New York Times*, May 11, 2011, in which he also notes his own proposal of the term "Anthrocene" in 1992. "Anthropocene" was later popularized by Paul Crutzen with Stoermer: see Crutzen, Paul J. and Eugene F. Stoermer. "The Anthropocene." *IGBP Newsletter*, 41. 2000: 17–18. There were earlier versions of this epoch, however different the term or its context: Comte de Buffon (1788), George Perkins Marsh (1864), Antonio Stoppani (1873), Robert Sherlock (1922), and Vladimir Vernadsky (1945) – see "Anthropocene" in *Keywords for Environmental Studies*, eds. Joni Adamson, William A. Gleason, and David N. Pellow (2016).

2 In conversation, 2016.

3 Of the Subcommission on Quaternary Stratigraphy, part of the International Commission on Stratigraphy.

4 The phenomena associated with the term include "an order-of-magnitude increase in erosion and sediment transport associated with urbanization and agriculture; marked and abrupt anthropogenic perturbations of the cycles of elements such as carbon, nitrogen, phosphorus and various metals together with new chemical compounds; environmental

changes generated by these perturbations, including global warming, sea-level rise, ocean acidification and spreading oceanic 'dead zones'; rapid changes in the biosphere both on land and in the sea, as a result of habitat loss, predation, explosion of domestic animal populations and species invasions; and the proliferation and global dispersion of many new 'minerals' and 'rocks' including concrete, fly ash and plastics, and the myriad 'technofossils' produced from these and other materials." http://quaternary.stratigraphy.org/working-groups/anthropocene/

5 Quoted from Hannah Devlin's "Was 1610 the beginning of a new human epoch?" *The Guardian*, March 11, 2015. www.theguardian.com/science/2015/mar/11/was-1610-the-beginning-of-a-new-human-epoch-anthropocene. See their 2015 paper, "Defining the Anthropocene" in *Nature* volume 519, 171–80.

6 Quoted in Robinson Meyer's "The Cataclysmic Break That (Maybe) Occurred in 1950" in *The Atlantic*, April 16, 2019. www.theatlantic.com/science/archive/2019/04/great-debate-over-when-anthropocene-started/587194/. The same article includes other examples, such as this one: "While Ruddiman's hypothesis is not widely accepted, it is taken seriously by the community. And his broader skepticism of codifying a late Anthropocene is shared by several members of the working group. In a separate paper published last week, five members of the committee rejected the idea of the 1950s Anthropocene. Today's scientists are simply too close to the events at hand to place a division in geological time, they argue. We don't yet know how significantly the planet's climate will change in the centuries to come: Will the shift be of the same magnitude as what occurred at the end of the last Ice Age, 12,000 years ago? Will it be equal to the first time that ice seized the surface of Earth, 2.1 million years ago? Or does it signal something far larger, a cataclysm on par with the asteroid impact that ended the dinosaur-dominated Mesozoic Era, 66 million years ago? 'There is no testable way of knowing at present,' they wrote." See also Ruddiman's articles in *Progress in Physical Geography: Earth and Environment* in 2018 and 2019.

The five authors also point out that the last 12,000 years would be understood as a single geological instant if they had happened millions of years ago.

7 See Sophie Moore's (et al.) "The Plantationocene and Plantation Legacies Today." *Edge Effects* 2019.

8 A description of the human-caused mass translocation of species, this term might also be used to describe what will remain after the Sixth Mass Extinction. For its original use, see Michael Samways's "Translocating Fauna to Foreign Lands: Here Comes the Homogenocene." *Journal of Insect Conservation* 3.2 (1999): 65–66.

9 Andreas Malm's and Jason Moore's term. See Donna Haraway's footnote number 6 in "Anthropocene, Capitalocene, Plantationocene, Chthulucene: Making Kin," *Environmental Humanities*, vol. 6, 2015, pp. 159–65.

10 See Kathryn Schulz's "What Is Distant Reading," *The New York Times*, June 24, 2011. www.nytimes.com/2011/06/26/books/review/the-mechanic-muse-what-is-distant-reading.html

11 Quoted from "UN Report: Nature's Dangerous Decline 'Unprecedented'; Species Extinction Rates 'Accelerating.'" www.un.org/sustainabledevelopment/blog/2019/05/nature-decline-unprecedented-report/ . For the full report, see the IPBES site: https://ipbes.net/

12 "The average abundance of native species in most major land-based habitats has fallen by at least 20%, mostly since 1900. More than 40% of amphibian species, almost 33% of reef-forming corals and more than a third of all marine mammals are threatened. The picture is less clear for insect species, but available evidence supports a tentative estimate of 10% being threatened. At least 680 vertebrate species had been driven to extinction since the 16th century and more than 9% of all domesticated breeds of mammals used for food and agriculture had become extinct by 2016, with at least 1,000 more breeds still threatened" (Ibid.).

13 "Shine, Perishing Republic" line 7.

14 "So uniquely vulnerable are the young that the World Health Organization estimates that children younger than 5 bear more than 40 percent of the global burden of disease caused by environmental risk factors and 88 percent of the disease burden caused by climate change." Frederica Perera, "The Womb Is No Protection From Toxic Chemicals," *The New York Times*, June 1, 2017. www.nytimes.com/2017/06/01/opinion/toxic-chemicals-pregnancy-fetus.html

15 For example, see www.pbs.org/newshour/show/the-dangerous-global-flood-of-misinformation-surrounding-covid-19 and www.washingtonpost.com/health/2020/02/10/coronavirus-is-spreading-rapidly-so-is-misinformation-about-it/

Works cited

Auden, W. H. *Selected Poetry of W. H. Auden*. Modern Library, 1958.

Calvino, Italo. *Six Memos for the Next Millennium*. Vintage, 1993.

Clark, Tim. *Ecocriticism on the Edge: The Anthropocene as a Threshold Concept*. Bloomsbury, 2015.

Devlin, Hannah. "Was 1610 the Beginning of a New Human Epoch?" *The Guardian*. March 11, 2015.

DuBois, Andrew. "Close Reading: An Introduction." *Close Reading*. Eds. Frank Lentricchia and Andrew DuBois. Duke, 2003. 1–40.

Jeffers, Robinson. *The Selected Poetry of Robinson Jeffers*. Ed. Tim Hunt. Stanford, 2001.

Kundera, Milan. *The Unbearable Lightness of Being*. Trans. Michael Henry Heim. Harper Perennial Classics, 1999.

Meyer, Robinson. "The Cataclysmic Break that (Maybe) Occurred in 1950." *The Atlantic*. April 16, 2019.

McFarland, Sarah E. *Ecocollapse Fiction and Cultures of Human Extinction*. Bloomsbury, 2021.

Parham, John. "Introduction." *Cambridge Companion to Literature and the Anthropocene*. Ed. John Parham. Cambridge UP, 2021. 1–39.

Perera, Frederica. "The Womb Is No Protection From Toxic Chemicals." *The New York Times*, June 1, 2017.

"Results of binding vote by AWG Released 21st May 2019." *Subcommission on Quaternary Stratigraphy Working Group on the 'Anthropocene.'* http://quaternary.stratigraphy.org/working-groups/anthropocene/. Accessed 1 December 2020.

Revkin, Andrew. "Confronting the 'Anthropocene." *The New York Times*, May 11, 2011.

Snyder, Gary. *The Real Work: Interviews and Talks 1964–1979*. New Directions, 1980.

"UN Report: Nature's Dangerous Decline 'Unprecedented'; Species Extinction Rates 'Accelerating.'" *Welcome to the United Nations: Sustainable Development Goals*. www.un.org/sustainabledevelopment/blog/2019/05/nature-decline-unprecedented-report/ Accessed 1 December 2020.

Yusoff, Kathryn. *A Billion Black Anthropocenes or None*. University of Minnesota Press, 2019.

Zalasiewicz, Jan, et al. "Anthropocene." *Keywords for Environmental Studies*. Eds. Joni Adamson, William A. Gleason, and David N. Pellow. New York University Press, 2016.

1

INHABITING WORDS, INHABITING WORLDS

A case for pragmatist close reading

Amelia Marini

I woke one night to the sound of antlers. The clacking near the window – was it the settling of the house? A mouse in the wall? – grew more distinct as I rose. The sound was unknown; it came from outside and I opened the window to hear it better. Behind the house is a corridor of open land, frequented by deer. Not much was visible, but I could hear hooves in the grass, the rumble of animal breath, and the sound that had brought me out of sleep: antlers striking other antlers. This could only be sparring, the contact tentative rather than violent, like a child with rhythm sticks. It lasted no longer than five minutes. Our lives are full of sound, more than is good for us, and little of it enchants – little of it sings – as do antlers in the night.

This essay is not about the California mule deer, but it cannot do without their song. I worry that we only attend to the song of things or compose our own songs about them once they are gone. Song becomes elegiac – a way of memorializing the things we have lost. As it is, we have lost a great deal, and stand to lose more. The title of this volume bears reference to this period of loss, the Anthropocene, a controversial but still widely used name for the era of human-caused catastrophe in which we find ourselves. However, as Donna Haraway cautions in *Staying with the Trouble: Making Kin in the Chthulucene*, the names we bestow upon this moment exert influence on the way we think about what we have done, what we can do, and who, beyond the human, might be involved. "Anthropocene," Haraway dissents, promotes a framework for the future – for solutions and action – that is not different in essence from the limited thinking that got us into the "trouble." Names are stories, she suggests, and the story that comes with *Anthropocene* does not inspire the kinds of the new, expansive, and inclusive narratives that we need in order to move beyond the partial tools that can no longer serve us (47). Nor does it, with its ties to capitalism, help us to return to the narratives of Indigenous peoples and cultures: those native to this land whose stories and songs about the earth offer richer frameworks for relational living than those offered by the Anthropocene.

This essay is based upon the premise that climate wisdom might indeed be a matter of learning to inhabit words and the worlds we compose with them. Within the Anthropocene, we've forgotten that songs and stories, in addition to eulogizing, have long been used to call forms of life into being. "Wise men," Ralph Waldo Emerson writes, "pierce this rotten diction and fasten words again to visible things" (23). When so much of the visible world seems to be rotting away – forests, sea ice, vast populations of animals – our rotting words, words that do not convince, describe, or enchant, may seem of secondary, rather than primary, concern. We forget that seeing and saying are looped processes: we do not only say what we see, we also tend to see the concepts we've learned to say. In the pages that follow, I will argue that the study of literature, particularly through the skill of close reading, offers a practice of creative attentiveness instrumental to the creation of livable futures on a "damaged planet" (Tsing M2). Close reading, as I will discuss, is more than a practice of decoding and demystifying literary ambiguities. Rather, it teaches us to better understand the connection between seeing and saying and presents an opportunity to inhabit texts in individual and personally resonant ways that require intersubjective imagining and relational thinking: the very skills that we must learn in order to live well upon this planet. In particular, this essay develops the concept of a "pragmatist" close reading – one inspired by the writings of the American philosopher, William James – as an essential tool for breaking down the habits of language and of thought that prevent us from realizing better narratives, and more livable futures.

My thoughts on the language of climate and the climate of language came into focus in the winter of 2018, with a composition course I taught at California Polytechnic State University called "The Rhetoric of Climate Change." Up to this point, I would not have called myself a nature lover and no one could have described me as any kind of outdoorswoman. However, the previous summer and fall I had found myself, for the first time, neck deep in articles, essays, and books about climate devastation. I read with appetite and agony. Strangely, a shift occurred. My body began to delight in sights and sounds I had previously taken for granted: the green smell of a tomato plant, the vibrational hovering of a territorial hummingbird, and the thickness of new spring growth on an oak tree. I sat for hours outside, examining the leaves of sage plants. In short, I began to see the natural world in my neighborhood with new vividness. The experience made me wonder if I could offer similar sensory awakenings to my students, or to other people who, like me, felt themselves on the periphery of the natural world. Could I induce sight, connection, and pleasure through the study of climate language? With this in mind, I planned the course.

I approached our first meeting with candor. I'm not an environmental scientist; I'm a scholar of literature and language. The seminar, designed so that I could teach responsibly, would attend to the ways that language shapes our understanding of and response to global climate crisis. Over that academic quarter, we explored the rhetoric of climate crisis as it is used by economists, environmental and political scientists, activists, spokespersons for industries such as agriculture and tech, liberal and conservative politicians, Indigenous leaders and thinkers, and literary writers. On the whole, the course was well-received by my students who engaged the material with generosity and insight.

However, one afternoon right before I went in to teach, a colleague stopped by my office. He asked how the class was going, and we talked for a few minutes before he sighed and said something along the lines of, "Yes, with everything one reads about climate change denial and global warming, it makes studying and teaching literature feel pretty useless." I'm sure I must have smiled and shrugged, letting the remark pass, but I thought about it for several days afterwards. I understood that the gesture was intended to be sympathetic, but the idea rankled. Especially on a warming planet, I did *not* feel that studying literature was a useless undertaking . . . but it was difficult to say why. Why study literature in an age of mass-extinction? Why study literature as coral reefs bleach, glaciers recede, and global emissions continue to increase? These are not new questions. They are simply the latest iteration of charges often leveled against the humanities' relevance. The difference is the implication that it is a moral failing to choose to study literature or philosophy, instead of merely an economic one. And, while the economic myth that English majors are unemployable needs no further debunking, it seems to me that we could spend a bit more time thinking through how the planet might actually benefit from having more students of literature, more students of story and song.

In fact, it may be very useful to begin to conceive of climate change denial as primarily a problem for the humanities. One of the more interesting findings of the climate course was that denial is not a problem of information. Among the first things the class read were numerous studies that show that there is almost *no* correlation between increased information about climate change and increased belief in its reality. Levels of climate change denial are not reduced by simply making climate data more widely available, or by creating more opportunities for education. Nor are rates of denial altered by more targeted "branding." Studies suggest that more persuasive messaging aimed at climate deniers – for example, studies that stress the need to move to renewable energy sources in order to increase national security, or to become more self-reliant – also fails to mitigate denial. Ultimately, the reason for this comes down to the increasing political polarization of the last 20 years. One study even suggests that ideological allegiance to partisan politics is currently so powerful that it prevails over all other factors. Individuals will vote the party line despite any and all evidence to the contrary, even if they find that evidence personally concerning or compelling (Dunlap 16).

Here, however, an interesting exception emerges. Despite the obstacle to climate action posed by party loyalty, one factor seems to demonstrate a modest ability to transform denial into conviction: personal experience of environmental changes. Rather than accepting information presented by others, farmers who watch their growing seasons shift well beyond the normal range, firemen who have felt the intensity of the last years' fires out west, divers and fishermen who can tell that our aquatic ecosystems are not well, begin to profess a greater belief that what we're experiencing is more than climate variability. Of course, to speak of a "personal experience" of climate change is to misunderstand the scale at which climate is measured. The environmental events that we experience on a daily, monthly, and yearly basis are weather events, not climate events. However, when sufficient data accrues to suggest that extreme weather seems to be becoming a normal, frequent occurrence, scientists can begin to measure it – and make predictions – within the greater context of global climate. And, on our end, the personal experience of change, of these yearly fires, floods, and droughts, begins to make us wonder, as human civilizations have always wondered in the face of great loss, if we haven't done something wrong.

In her memoir, *Thus Spoke the Plant*, the plant biologist Monica Gagliano blames "faulty thinking" that must be revised in order to allay the destruction (101). In particular, she faults the kind of thinking that, within the professional sciences, has maintained a mechanistic and anthropocentric perspective of the nonhuman world that originates in Aristotelian classification. "By defining insensitivity as the key criterion to differentiate plants from animals," she writes, ". . . the Aristotelian story had, in effect, transformed plants into objects – a spurious idea that still sanctions the human right to use (and abuse) plants and that exempts us from any sense of the responsibility or respect toward them as living beings" (105). Though Gagliano's critique is primarily aimed at scientific institutions and the narrow forms of knowing that constitute their methods, it's easy to see objectification reflected within our culture at large, especially as it shapes American consumer practices. As Bill McKibben suggests in the opening of his book-length argument against economic growth, "For most of human history, the two birds More and Better roosted on the same branch. . . . It is no wonder and no accident that they dominate our politics, our outlook, even our personalities" (1). As McKibben argues, our cultural devotion to measuring individual success in terms of accruing ownership and wealth – the belief that we are only worth what we can buy – is not only devastating for our planet which is plundered to produce all those things and becomes a wasteland of all we cast off, but also leads to despair. Materially, Americans have more than they've ever had before, and have never reported greater levels of anxiety and unhappiness (38).

If we're to take Gagliano's critique seriously, it becomes clear that we will need new methods for thinking through the crisis that confront us, methods that decenter "profit" and "progress" from our narratives and refuse to reduce the living world to a machine or a set of "resources." As a corrective to unimaginative and objectivistic thinking, the practice of literary studies can offer a great deal. The task of literature, to return to Emerson's *Nature*, is to fasten language to life, and life to language. One model of this might look like the project undertaken by Robert MacFarlane and Jackie Morris called *The Lost Words*, a beautifully illustrated book of acrostic "counter-spells" comprised of nature words that have been dropped from the *Oxford Junior Dictionary*. The *OJD*, an abridged children's dictionary facing pressure to add more technologically relevant words such as "broadband" and "chatroom," made room for these additions by dropping other presumably less relevant words, a disproportionate number of which pertain to the natural world like "kingfisher" and "fern." MacFarlane and Morris's project attempts to save these words, and the creatures they name, from erasure. One response to the project expresses the urgency of the work particularly well:

> If we are to imagine the complex solutions that our continued existence on Earth demands, we cannot conceive of the environment as a green-brown blur, of nebulous and inarticulable importance. . . . Perhaps, in the face of leaders who respond to environmental catastrophe by gutting environmental protections, a vocabulary of enchantment and love, as opposed to anger and fear, is something to heed. A vocabulary that helps us identify, with clarity and spark and wonder, what we hold dear and what we thus need to fight for.
>
> *(Sharma)*

A related project is undertaken by *Counter-Desecrations: A Glossary for Writing Within the Anthropocene*, a collaborative collection of reimagined definitions that help us to "question, break-open, and reorganize thinking" (12). Both projects suggest that it's difficult to save something we do not know intimately, and it's difficult to know a thing intimately without naming it, perhaps over and over, in millions of different ways. Each of these names helps us to see the world as the complex, infinitely faceted entity that it is. A favorite passage from William James's *Pragmatism* lists the varying names of a certain constellation: "Charles's Wain, the Great Bear, or the Dipper. None of the names will be false, and one will be as true as another, for all are applicable" (597). James calls our new names for old things "additions" and suggests that to add to the world in this way is a primary function of the human being. "The great question is," James suggests, "does [the world], with our additions, *rise or fall in value?* Are the additions *worthy* or *unworthy*?" (598). The living, agential world, as it exists in reciprocity with human beings, requires a living, hence literary, language: a language capable of interpretation, flexibility, openness, and change.

Ours, after all, is not the only discipline to name things. Taxonomy depends upon the classification of organisms into named branches. But the study of literature is the only discipline in which the mystery and ambiguity of naming a thing is as celebrated a function of language as is elucidation or demarcation. In literature, things are named in order that we might know the truth of their mystery more deeply. If literary language can be said to illuminate, it does so only in a flickering half-light. Recall the butterfly in the preface to Nathaniel Hawthorne's *House of the Seven Gables*. Rather than "impaling" its subject "with an iron rod, or rather, as by sticking a pin through a butterfly – thus at once depriving it of life, and causing it to stiffen in an ungainly and unnatural attitude" literary language depends upon "a far more subtle process" (352), one that *involves* the reader in the complex ambiguities of meaning-making.

This "involvement" in ambiguity is the antidote that literary studies offers to a world made sick by consumer practices. Literary experiences are differentiated from the act of reading for information or for entertainment in that they are not situations of passive consumption. So often the act of reading amounts only to a polite and complacent "keeping up with culture" – even if the works themselves are extraordinary (Lorentzen). We finish a work, and it moves through us, leaving no trace. Literary experiences, on the other hand, mark us. They are sites of relationship-building in which we use language to entangle ourselves more deeply in the shared human labor of connection and questioning, of thinking and expression. Works of writing become literary when they are treated, actively, as literature. This requires the reader to meet the text with energy and ideas – to engage it – in order to participate in the meaning of the text. Sometimes literary encounters are pleasurable, other times not; we know that the act of struggling or wrestling with a challenging text often provides more illumination than those instances when we feel ourselves entirely at home in the work. However, whether or not she enjoys reading a work, the reader bears some responsibility for the literary meaning that emerges from a text; she is implicated in it. As science fiction writer Ursula K. Le Guin writes,

> Science describes accurately from outside; poetry describes accurately from inside. Science explicates; poetry implicates. Both celebrate what they describe. We need the language of both science and poetry to save us from merely stockpiling endless 'information' that fails to inform our ignorance or our irresponsibility.
>
> *(Tsing M16)*

To return to the finding of my class, our encounters with information – with external material, data, "the text" – are not sufficient to prompt the kinds of widespread action that's required of us. We need implicating, experiential encounters as well – data contextualized within the familiar, intimate field of our everyday experiences: the text *in relation* with an individual reader. Though my class was on the rhetoric of climate change, what I've learned is that perhaps we really need a poesis of climate change: a way to think about how the reader or receiver of "information" *makes* something from it. Furthermore, a poesis of climate change would suggest that this meaning-making can and should be personally demanding, creatively open, and imaginatively rich. This is the work of literature.

Unfortunately, the English classroom is not always a richly imaginative space. It's even possible that the sciences have pedagogically surpassed the humanities in their ability to create hands-on, open-ended, self-directed encounters with their subject-material. Perhaps readers of this essay teach, as I do, students who understand literary reading to be a kind of tedious decoding. Many students come to us believing that the author of a text intends a meaning that exists just under the dialogue and events of the text. They insist that if we could only crack open the character of Hester Prynne, for example, we would see that she stands for feminism, or Calvinism, or repression. Among students who feel that it's worth their time to decipher these riddles which seem to have nothing to do with them, it takes a great deal of work to pry these orthodoxies out of their readings. However, if we do, we might be able to present them with the possibility that ideas that defy orthodoxy, those flickering readings that won't be tidily contained into boxes, are often more interesting, fruitful, and pleasurable to inhabit.

And, we must be honest; we know that students are not the only ones with attachment to orthodoxy. How often do we in the field read with our preferred "-isms" at the forefront of our thinking? Materialism, historicism, Marxism, psychoanalytic criticism: we create professional identities by speaking of texts in ways that are codified. However, the codes, useful as they sometimes are, have come to *signal* identity as much as anything else. Fluency in particular forms of discourse grants the feeling of insider-status that the institutions – with their increasing reliance upon insecure and undercompensated labor – have failed to deliver. In many ways, I think that new entering scholars like myself, especially those who seek traditional academic employment, compensate for the terrible insecurity of academic labor by clinging ever more tightly to these codified markers of scholarly identity. Illegibility is too great a risk.

Tragically, the creative energy and intellectual flexibility of my peers are exactly what is needed to propel away from orthodoxy and toward experimental and collaborative discourses that extend into other disciplines, particularly into those fertile places where the humanities meet the sciences. Interdisciplinary niches such as new materialism, ecocriticism, and science studies – especially when informed by an interest in aesthetics – have been instrumental in making this move and creating supportive spaces for experimental scholarship. However, it's essential that more of these transdisciplinary conversations and creative encounters occur. What we are learning across all disciplines in the wake of climate change is that the orthodoxies of the past do not equip us to well-perceive the terrain that we're moving through. Just as evolutionary biologists begin to articulate the limits of a discourse that focuses predominantly upon individual fitness and competition, so we must also ask whether our allegiance to the "-isms" of our discourse has prevented us from seeing our own literary terrain in open and original ways: ways that might better facilitate personally implicating, poetic encounters with language. Or, as Emerson says, ways that might "fasten words again to visible things."

Close reading, a method championed by New Critics such as Cleanth Brooks in the 1930s and 1940s and still used, at least in name, by many if not most English classrooms today, was intended to do exactly this: to permit readers to engage the text according to nothing but the life and logic of the text itself. Rebelling against what Brooks considered to be a tyrannical emphasis upon the biography of the author, or the historical context and concerns of a text's emergence, the close reading method of the New Critics suggested that everything that one needed to know about the poem could be found within the language and structure of the poem. The text, within this methodology, exists as a complete and perfectly unified "autotelic" world of signification, even when elements seem contradictory or incongruous. It is not surprising that this approach is still so widely used in contemporary classrooms. In theory, close reading gives all students an equal footing. It requires no outside knowledge or research; close reading is a matter of the reader and the text in front of her. However, the problem with new-critical close reading is that the emphasis placed on a complete and unified text is precisely what leads so many students to feel that texts are daunting, monolithic entities to be decoded – that the elusive unity must be found or figured out. Brooks himself describes form as the "solution" to a "problem": "The structure obviously is everywhere conditioned by the nature of the material which goes into the poem. The nature of the material sets the problem to be solved and the solution is the ordering of the material" (Leitch 1218).

Taken thusly, literature is not the open-ended experience that occurs when the reader encounters a text; instead, it is a problem the reader must solve with a limited number of tools. We see how disastrously this plays out in the modern classroom. Students, intimidated by "great" works feel themselves inadequate to the task of reading "on tiptoe," as Henry David Thoreau says (99). Since the work stands alone – great, unified, and untouchable – students may wonder why it matters whether they read it or not. Especially if only one unified meaning inheres within a work, why can't someone else do the work of finding it? Why should *I* struggle with a work only to see what a lot of other people have already seen? I suspect this is the reason why study guides like CliffsNotes, SparkNotes, and others are still so popular. The text has *a* meaning, and in order to pass the particular bar the text presents, the reader must know the code.

Happily, this isn't always how modern close reading plays out. The backlash against the New Critics by deconstructionists, reader-response critics, feminist scholars, and new historicists who insisted that the individual reader mattered and that the context of history was illuminating rather than inhibiting, has helped close reading to evolve into a more diverse practice. While the emphasis remains on creating interpretations of a text through a deeply focused attention to language, the method has become more open to both a reader's own experience and identity, and outside contextual information and ideas. I myself practice a close reading that's influenced by William James's pluralistic pragmatism – which allows borrowing-from and rejecting all "isms" as needed – and find his descriptions of the difference between rationalism and pragmatism helpful for distinguishing between a new critical approach to close reading and a pragmatist one. "The essential contrast," James writes, "is that *for rationalism reality is ready-made and complete from all eternity, while for pragmatism it is still in the making, and awaits part of its completion from the future.* On the one side the universe is absolutely secure, on the other it is still pursuing its adventures" (599). If we apply this concept to the question of literature, we might phrase it in this way:

a rationalist, or new-critical, approach to a text believes that the text itself *is* literature. The pragmatist approach, on the other hand, believes that literature is something that occurs in the interactions between the text and the world. The text is new for each reader, and literature is something that the text *does*, rather than something that the text *is*.

To approach literature as a pragmatist gives it new urgency. Why read a particular text? Because each person who engages the text is the only person who can bring her particular literary experience into being. A text will never speak exactly to another in the way that it speaks to you. Refusing to engage a text is to deny an original form of experience from coming to be. This doesn't mean that all readers are well-prepared to read all texts, or that all literary experiences will be of equal value. Sometimes we encounter a work too early, or too late, or in a state of mind that is otherwise inconducive to literary experience. To return to an example from Nathanial Hawthorne, this is the reason why so very many high school students abhor *The Scarlet Letter*, a truly incandescent work. Who you are at the time that you read a text is important; not all teenagers have had the experiences of Eros, the emotional maturity to imagine those experiences, or the patience to give themselves over to the subtlety that the novel demands. Reading the *Scarlet Letter* isn't simply a matter of being able to regurgitate the novel's themes, or of interpreting the intentionally ambiguous "A". To read the text is to be in relationship with it, to bring oneself to its pages and to see not what one is "supposed" to see there according to CliffsNotes or Harold Bloom, but to see the things that one *actually does*. This is the reason why literary works not only allow for but demand rereading. As we change, so do the works.

The distinction between literature as something that exists unto itself, and literature as an experience of being-in-relation is a crucial one to understand if literary studies has a part to play in facilitating implicated relationships with the more-than-human world. If belief in climate crisis hinges upon an individual's ability to perceive her relationship to her environment, it matters whether or not we teach literature as a practice of being in relation, or whether we approach texts as things that stand apart from us. When we teach literature, we need to be teaching more than the names, dates, and themes of various authors. Instead, we need to be teaching a kind of perception that emphasizes connectivity, complex communication, and the interweaving of many types of experience: a kind of perception with value beyond the literature classroom. To do so will require that we change the way we approach close reading.

Often when we teach students how to close read, we model how to work at the level of the sentence, spinning a word, an idea, or a motif into an interpretation. These skills of focus and attentiveness are valuable; however, attention to the text only makes up half of the experience of close reading. What we need to bring to light is the idea that close reading is not simply a practice of being close to the text, but of *allowing the text to be close to us*. Close reading is just as much a matter of the text reading me, as it is one of me reading the text. After all, unless I ask my students to do a close reading in the "style" of some other person or party – a Freudian reading or a Marxist reading, for example – it will be their own concerns, affinities, and aversions that materialize. I think we're not being quite honest if we don't allow that when we're close reading, we're doing something personal. We bring all of our prior experiences, traumas, intelligences, esthetics, desires, fears, preoccupations, and interests to the text. This, again, is why timing matters. As our fears or preoccupations change, so do our abilities to read certain works. As the pragmatist critic Richard Poirier writes, "Reading is nothing if it is not personal. It ought to get down ultimately to a struggle between what you want to make of a text and what it wants to make of itself and of you" (167).

Close reading is an invitation to invest oneself in a poem, novel, or essay, and to realize that meaning doesn't stand alone within a text but emerges through a collaboration of minds. I don't think that there's any one way to encourage such readings, but we might start by inviting our students to ask questions like "What am I seeing? What in my experience up to now has primed me to see this? What can I say about this that someone else might not be able to say? What questions emerge for me? How does this passage make me think about my own experiences differently?" Of course, the inevitable critique of such questions raises concerns about whether they invite a kind of parochialism or even narcissism into the classroom. Am I suggesting that students must be able to "relate to" a work of literature in order to have an experience with it, and isn't this isolating? To what extent does this practice of close reading make room for texts that challenge students' experiences? These are important questions.

Indeed, I am in some ways suggesting that students must be able to "relate to" a piece of literature, but "relation" here means more than "does this character look, think, act like me?" or "do I agree with what I perceive the author to be saying?" When I suggest that students need to be able to relate to a work to read it well, I consider relation to be a function of imagination, and I'm asserting that the building of strong imaginative skills requires intention, practice, openness, and empathy. An unimaginative student is going to have a difficult time reading works they cannot "relate" to, while a student of greater imaginative flexibility will be able to gather much from a text that very little resembles her own day-to-day life. I cannot say this better than Zadie Smith does in her recent "Defense of Fiction":

> I lived in [books] and felt them live in me. I felt I *was* Jane Eyre and Celie and Mr. Biswas and David Copperfield. Our autobiographical coordinates rarely matched. . . . But I'd been sad and lost, sometimes desperate, often confused. It was on the basis of such flimsy emotional clues that I found myself feeling with these imaginary strangers: feeling with them, for them, along them and through them, extrapolating from my own emotions. . . . The voices of characters joined the ranks of all the other voices inside me, serving to make the idea of my "own voice" indistinct. Or maybe it's better

> to say: I've never believed myself to have a voice entirely separate from the many voices I hear, read, and internalize every day.
>
> *(Smith)*

The experience of intersubjectivity that Smith points to, the kinds that literary experience demand, is precisely those that are needed in our interactions with the natural world in order to attune ourselves better to the health and function of ecosystems. I'm struck by the resonances between the last sentences of Smith's reflection, and these by Gagliano:

> How can a plant readily know us when we are hardly aware of the plantness within ourselves? . . . Perceived through the veil of motionlessness even by our most brilliant minds, this plantness has seemed inaccessible and even deliberately hidden from us. . . . When we feel it, this plantness . . . [what] appeared to be intangible – an obscure otherness – is unveiled as the intimate familiarity of an obvious and luminous likeness.
>
> *(16)*

The comparison of the two quotations prompts the following question: is intellectual, emotional, or aesthetic resonance with an imagined person through the particularities of language such a far stretch from experiencing similar connectivity with embodied, nonhuman forms of life through the particularities of ecosystems? While imaginative, intersubjective thinking is permissible, encouraged, and taught within the realms of literary experience, it still struggles to find footing as a valid component of scientific inquiry and public discourse. It seems that we have not yet recognized the importance of "feeling with" the beyond-human world as a strategy for prompting scientific inquiry, national policy, or political mobilization.

If we did, we could look to a long history of indigenous ecological practices, in which intersubjective and relational thinking has always played an important role. Robin Wall Kimmerer, a botanist and member of the Citizen Potawatomi Nation, laments how long it has taken for western scientists to accept – or even to regard with possibility – what Native people have long known about the behaviors, needs, and balances of their local ecologies using indigenous epistemologies. Kimmerer cites tree communication as a prime example. "In the old times, our elders say, the trees talked to each other. . . . But scientists decided long ago that plants were deaf and mute, locked in isolation without communication" (19). Now, western science has demonstrated that trees do indeed communicate through their pheromones and through vast underground mycorrhizal networks. Trees warn one another of environmental threats and share nutrients between themselves in times of need (20). There can be no question that there's a kind of community sentience in play, even if the communication mechanisms differ from our own. This is not to caricature or romanticize indigenous ecological practices as an ideal of perfect stewardship, as scholars such as Astrid Ulloa have warned against (131). However, there is much to be learned from an epistemological approach that is not based on utility or objectification. "The questions scientists raised," Kimmerer writes, "were not 'Who are you?' but 'What is it?' No one asked plants, 'What can you tell us?' The primary question always was 'How does it work?' The botany I was taught was reductionist, mechanistic, and strictly objective. Plants were reduced to objects; they were not subjects" (42).

The difference between those questions, "Who are you?" and "What is it?" cannot be overstated. The latter opens an investigation while the former invites an exchange. Investigations are controlled, anticipated, and unilateral. One party decides how and when to probe, act, experiment, and observe while the other passively endures the inquiry. However, when we're invited to participate in an exchange, we give up some of our autonomy in order to better facilitate relationships. Exchange is conversational – it exists only if neither party asserts dominance and if both are willing to play active and receptive roles as they move toward some as-yet-to-be-determined end (Ingold 34). The misconception is that investigation is rigorous and objective, whereas exchange is sentimental and unprofessional. Nor are these prejudices limited to the sciences. As the literary theorist Eve Kosofsky Sedgwick noticed, literary study is too often undertaken in a spirit of investigation and exposure rather than conversation for the sake of an imagined standard of "rigor" (138). Since Sedgwick's observations, a "post-critical" movement of scholars has advised wariness against literary discourse that embraces objectivizing terms like "investigate", "interrogate", and "examine" at the expense of other less domineering discourses (Felski 17). In literary studies, we are afforded the space to recognize the way that these discourses mirror colonialist practices of conquering territory and imposing an outside order instead of attuning ourselves to the richness of an already present order. To shift away from a critique based in objectifying language could mean that close reading, rather than an experience of reading against the grain of the text, could be experienced as a process of becoming more conversational with and more deeply attuned to the text.

Just as people were once attuned to land. The more I insist upon attunement as a part of my own disciplinary practice, the more I feel myself drawn into something older than the written word. Literary studies, far from inventing deep attunement – call it close reading or something else – is merely one space in which this kind of attention could be practiced with an eye toward an ecological context. As Barry Lopez writes in *Arctic Dreams*, "The land is like poetry: it is inexplicably coherent, it is transcendent in its meaning, and it has the power to elevate a consideration of human life" (274). Lopez's invocation of an "inexplicable coherence" resonates deeply with the way in which the New Critics spoke of literature. The one revision I might suggest however, is that if literary studies is to participate in the shifting of consciousness that climate crisis requires and if words are to become attached to visible things, we may need to invert Lopez's simile. Perhaps it is poetry that aspires to the qualities of landscape. To position attunement as a primary goal of close reading might help us to understand that land was the original literature, and land the being to which literature could help us return.

To do so will require an engagement that is nothing short of immersive. As with language, all landscapes have a grammar of their own. To this point, Robin Wall Kimmerer links indigeneity to a kind of fluency: "To be native to a place we must learn to speak its language" (48). Ecologically or literarily, it is difficult to read deeply if we're working within a limited vocabulary. It is not enough to know several individual parts: a verb, a species, a noun, a mineral. Languages and ecologies work as systems in which each individual part moderates the activities and functions of all the other parts. We need to understand these parts, but we also have to be willing to shift our vision from part to whole. Like a concert pianist for whom individual notes have become melody, we would ideally read in such a way that we could keep both parts and wholes in consciousness at once. In a passage describing the immersive attention of Inuit hunters, Barry Lopez re-creates this unity of dual focus. Lopez writes:

> Hunting . . . is a state of mind. All of one's faculties are brought to bear in an effort to become fully incorporated into the landscape. It is more than listening for animals or watching for hoofprints or a shift in the weather. It is more than an analysis of what one *senses*. To hunt means to have the land around you like clothing. To engage in a wordless dialogue with it, one so absorbing that you cease to talk with your human companions. It means to release yourself from rational images of what something "means" and to be concerned only that it "is." And then to recognize that things exist only insofar as they can be related to other things. These relationships – fresh drops of moisture on top of rocks at a river crossing and a raven's distant voice – become patterns. The patterns are always in motion.
>
> *(200)*

This passage offers a profound depiction of reading by emphasizing the impossibility of artificially isolating any particular "part" from the whole of its environment. The hunter doesn't consciously note environmental conditions and their corresponding consequences; he enters into the flow of many moving parts. Still, we must not mistake this flow, continuity and co-extensiveness with the "green-brown blur" warned against in *The Lost Words* response. Flow emerges not out of a lack of attention – the refusal to focus the rational mind on specifics – but through a meditative absorption. I suspect that it is only through repeated rational exercise, as practiced in all forms of language acquisition, that one ever enters into the state of living, almost effortlessly, within an environmental grammar. The result of concerted learning is, finally, a way of being. Flow also depends upon a relationship to time that is at odds with capitalist ideals of efficiency, maximization, and progress. In order to practice fluency, we must be open to a kind of hyper-attentive wandering. "[To] a non-Eskimo observer," Lopez writes, "the Eskimo might seem to have 'no sense of direction.' And because he travels somewhat like the arctic fox – turning aside to investigate something unusual, or moving ahead in a series of steps . . . instead of in a straight, relentless dash for the 'goal' – the Eskimo might be thought . . . improvident" (277). To the contrary, Lopez asserts, it is only when we free an environment of our "use" for it, that we come to know its characteristics more truly. In other words, in releasing ourselves from ends-based thinking, we are permitted to experience a landscape in much greater variety, attention, and depth.

The same can be said for texts. This is one of the great ironies and difficulties of teaching close reading within the modern high school or university. Our school systems are designed to measure progress through educational staging, grading, testing, and learning objectives. Our students always have a goal in mind, whether it's a top grade, or simply to progress in the sequence of their studies. And yet, to truly enter into a practice of close reading, we need to encourage them to wander, to flow – to give up the goal. Often, without entirely realizing it, we are asking for both things at once; we ask them to progress through a certain number of pages each week *and* to develop attentive, fluid close readings of those pages. Because, culturally, we place so much more emphasis on ends-based thinking, it is not surprising that our students feel less at home, even anxious, in situations where the goal is not clearly defined. In fact, the latter is considered poor pedagogy. This is a problem. If our students are always looking for what we ask them to look for, they're robbed of opportunities for original sight unmediated by the instructor's perspective. Of course, we should model for students what it could look like to wander inside the text – but the emphasis must remain on cultivating students' own confidence to roam, even if the early efforts yield less than we would hope. Ideally, as they practice, our students become like hawks circling a field, looking for movement. Eventually, they seize on the word or idea that jumps, the resonance with potential to become a close reading, and relationship all their own.

However, once we have located a site for reading, attending to it to learn something about both the text and ourselves, part of treating a text like an ecology, or an ecology like a text, is to remember that the work of literature is much more than any one person can say about it. "The land," Lopez writes, "retains an identity of its own, still deeper and more subtle than we can know. Our obligation toward it then becomes simple: to approach with an uncalculating mind, with an attitude of regard . . . to preserve some of the mystery within it as a kind of wisdom to be experienced, not questioned" (228). To maintain this balance between sharp perception and openness to mystery is a particularly literary tension. It is also humbling. Even though the syntax of a published work may be fixed, meaning flows through those structures like water through a canyon. And like a canyon wall eroded by a river, words display etymological traces of their histories, but they cannot hold water, or meaning, in a permanent way. Rather, they present an invitation to pour *ourselves* through their shapes, and to see how we emerge on the other side.

It is an intimate endeavor. Literature teaches us to see the details of the world more clearly by coming close, and in coming close we're offered a better sense of the world's immensity. To believe in the immensity of the world – in the immensity of language, of nature, of other living beings, and of our own individual and collective potential to do well by them – restores enchantment to a world that was never dead, except in our understanding of it. Words are an extension of the world we inhabit, an extension in which imagination, play, complexity and creativity are called upon to perform restorative work in a culture of objectification and isolation. It is time to allow the work of words to become our work in the world.

Works cited

Dunlap, Riley E, et al. "The Political Divide on Climate Change: Partisan Polarization Widens in the U.S." *Environment: Science and Policy for Sustainable Development*. 58:5, 4–23, 25 Aug. 2016.

Emerson, Ralph W, and Joel Porte. *Essays & Lectures*. Literary Classics of the U.S., 1983.

Felski, Rita. *The Limits of Critique*. Chicago, 2015.

Gagliano, Monica. *Thus Spoke the Plant: A Remarkable Journey of Groundbreaking Scientific Discoveries & Personal Encounters with Plants*. North Atlantic Books, 2018.

Haraway, Donna J. *Staying with the Trouble: Making Kin in the Chthulucene*. Duke, 2016.

Hawthorne, Nathaniel. *Novels*. New York: Literary Classics of the United States, 1983.

Ingold, Tim. *Anthropology and/as Education*. Routledge, 2018.

James, William. *Writings, 1902–1910*. Literary Classics of the United States, 1987.

Kimmerer, Robin Wall. *Braiding Sweetgrass: Indigenous Wisdom, Scientific Knowledge and the Teachings of Plants*. Milkweed, 2013.

Leitch, Vincent B. *The Norton Anthology of Theory and Criticism*. W.W. Norton, 2010.

Lopez, Barry H. *Arctic Dreams: Imagination and Desire in a Northern Landscape*. Vintage, 2001.

Lorentzen, Christian. "Like This or Die: The Fate of the Book Review in the Age of the Algorithm." *Harper's*, Apr. 2019, harpers.org/archive/2019/04/like-this-or-die. Accessed 15 May 2020.

Macfarlane, Robert, and Jackie Morris. *The Lost Words: A Spell Book*. Hamish Hamilton, 2018.

McKibben, Bill. *Deep Economy: The Wealth of Communities and the Durable Future*. St. Martin's Griffen, 2007.

Poirier, Richard. *Poetry and Pragmatism*. Harvard University Press, 1993.

Russo, Linda, and Marthe Reed. *Counter-Desecration: A Glossary for Writing within the Anthropocene*. Wesleyan, 2018.

Sedgwick, Eve K. *Touching Feeling: Affect, Pedagogy, Performativity*. Duke, 2006.

Sharma, Meara. "How the Loss of Vivid, Exacting Language Diminishes Our World." *The Washington Post*, 8 Dec. 2017.

Smith, Zadie. "Fascinated to Presume: In Defense of Fiction." *New York Review of Books*, 24 Oct. 2019.

Thoreau, Henry D, and Bill McKibben. *Walden*. Beacon, 2017.

Tsing, Anna, et al. *Arts of Living on a Damaged Planet*. University of Minnesota Press, 2013.

Ulloa, Astrid. *The Ecological Native: Indigenous Peoples' Movements and Eco-Governmentality in Columbia*. Routledge, 2005.

2

ECOPOETICS AND THE MYTH OF MOTIVATED FORM

Greg Garrard and Rina Garcia Chua

The study of English Literature in the mid- to late twentieth century, particularly in the classrooms of Britain and its former Empire, took its moral purpose from Leavisism but its method from IA Richards. The pedagogy and practice of close reading are derived from Richard's *Practical Criticism* (1929), which reported on experiments carried out at Cambridge on readers presented with poems stripped of both author and historical context. Richards claimed that readers in these circumstances encountered typical "difficulties" (Richards 13), from failure to understand the "plain sense" of a poem through the "Scylla and Charybdis" of "sentimentality" and "hardness of heart" (16) to distracting "doctrinal adhesions." Close reading – which Richards asserts "All respectable poetry invites" (203) – is intended to counteract these difficulties by bringing sustained attention to the words on the page. Since this method happens to align with the exigencies of the unseen written examination, the supposed gold standard of assessment in British and imperial education systems, practical criticism, and close reading took their place at heart of English Literature pedagogy throughout the remainder of the century.

One of Richards's difficulties is "sensuous apprehension," or the challenge some readers experience in sensing poetic rhythm: "The gulf is wide between a reader who naturally and immediately perceives this form and movement (by a conjunction of sensory, intellectual and emotional sagacity) and another reader, who either ignores it or has to build it up laboriously with fingercounting, table-tapping and the rest" (14). The fortunate sensitive reader, though, having accurately divined the rhythm and deduced the meter, might be inclined to ascribe *too much* significance to the pattern. Indeed, "not a few" of Richards's experimental cohort maintained that "the meaning of the words is irrelevant to the form of the verse" and that "this independent form possesses intrinsic merit" (232). Bringing his trademark humor and scientific skepticism to the question,

Richards tests this assumption by creating what he calls a metrically identical "double or dummy which at least comes recognizably near to possessing the same virtue." Comparison of the dummy with its original, Milton's *On the Morning of Christ's Nativity*, will "show how much the sense, syntax and feeling of verse may serve as an introduction to the form." From Richards's uncompromising observation that "stupidity is not a simple quality" and that the "most leaden-witted blockhead thereby becomes an object of interest," (43) this essay derives a principled curiosity about dummies who mistakenly think that poetic forms have meaning and purpose *independent of their semantic content*. We will be calling this error "the myth of motivated form."

Richards's target, in this instance, is a reader who attributed a "gentle sadness" to the rhythmic form of John Donne's sonnet "At the round earth's imagin'd corners," and concluded "I should never bother about the sense; the sound is enough for me" (49). "Commentator" 3.6's "protocol" (Richards's terms for his respondents and their papers) seems, in fact, the mildest form that a statement about poetic rhythm might take – its imprecision makes it irrefutable, no more. Here we assess stronger claims from ecopoetic theory about the significance of forms; contrast them with what we see as more modest and sustainable assertions; test them in relation to two North American ecopoems; and conclude by considering the implications of our critique for close reading as a pedagogical practice.

The myth of motivated eco-poetic form

One of the "landmarks in literary ecology" collected in *The Ecocriticism Reader* is William Rueckert's 1976 "Literature and Ecology: An Experiment in Ecocriticism." Demoralized by the churn of critical trends (even before the advent of High Theory!), Rueckert looks to the science of ecology for touchstones of moral import and creative systems-thinking that, he hopes, can account for the beauty and value of poetry. Having warned of the need to "avoid victimage or neutralization by simple-minded analogical thinking" (Glotfelty and Fromm 109) – an ambiguous advertisement that seems as much anticipatory excuse as acknowledgment of risk – Rueckert proposes that:

> Some poems – say *King Lear, Moby Dick, Song of Myself* – seem to be, in themselves, ever-living, inexhaustible sources of stored energy, whose relevance does not derive solely from their meaning, but from their capacity to remain active in any language to go on with the work of energy transfer, to continue to function as an energy pathway that sustains life and the human community.
>
> *(108)*

Where much of human activity is seen as entropic, burning and consuming in pursuit of transient, ignoble, or simply impossible ends – unending economic growth, for example – poems are portrayed as "negentropic," like green plants.

Though Rueckert says little explicitly about poetic form – not least because he uses "poem" broadly to refer to a range of influential literary texts – his account relies heavily on a half-articulated analogy between the structure of ecosystems (i.e., pathways of material, energy and information flow) and that of poems:

> What a poem is saying is probably always less important than what it is doing and how – in a deep sense – it coheres. Properly understood, poems can be studied as models for energy flow, community building, and ecosystems. The first Law of Ecology – that everything is connected to everything else – applies to poems as well as to nature.
>
> *(110)*

Since poetic form, not environmental themes, endows poetry with ecological significance, perhaps even scientific ecologists ought to pay attention.

Rueckert deserves credit for his risky and, as it later turned out, productive experiment. The problem is that, his own caveats notwithstanding, poems are really *very* unlike ecosystems. The discourse of "energy," which yokes together thermodynamics and human linguistic creativity, verges on New Age talk of crystals activating chakras. As Dana Phillips puts it with characteristic pungency:

> even if literary form and the form of ecosystems could be shown to be more similar than I suspect they are (which is highly unlikely, since no one knows whether ecosystems, not being organisms themselves, can be said to have forms), their similarity still would be only a coincidence, and not something we would be bound to regard as important.
>
> *(Phillips 144)*

Indeed, ecologists no longer operate on the assumption that ecosystems are non-arbitrary associations: "Examined on a long-term time scale, the various species that today form associations are really accidents of history" (Kricher 90). Not only are poems unlike ecosystems, ecosystems are also wholly unlike poems.

Angus Fletcher presents a far more detailed and developed account of motivated form in *A New Theory for American Poetry: Democracy, the Environment, and the Future of Imagination*. He gets off to a good start with a wonderfully nuanced and illuminating contrast of William Wordsworth and John Clare, though the claim that there is an "almost genetic connection between poetry and natural fact" (Fletcher 22) is alarming. Fletcher argues that "when Romantic poetry turned to its involvement with nature, it committed itself to deepening, analyzing, but generally idealizing a practice which the study of nature makes virtually unavoidable, namely, the description of the natural scene" (24). The pursuit of such description inevitably raised "the argument of form" (as indeed should any consideration of poetry whatsoever), and yet poets understood form in an opposite sense from Plato, as enabling change rather than embodying constancy. From this one might deduce that the dynamics of change in poetics operate wholly independent of nature, and Fletcher does acknowledge a degree of arbitrariness in his contrast of British topographic poetry and the American "Chorographic vision [that] questions topos or *place* by showing turbulent movements within *space*" (269). Once he begins to discuss the poetry of Walt Whitman, though, his enthusiasm overwhelms his discretion, and he makes a series of increasingly extraordinary assertions about the isomorphic relationship he perceives between Whitman's catalog poems, natural environments, and American democracy. First, Whitman's volleys of bombast are said to have recast "the American sentence" (100) from an old-world grammar of property ownership to individualistic "phrases" that encompass American diversity. Second, Whitman's poetic somehow also embodies or encapsulates "environments":

> His poems are not *about* the environment, whether natural or social. They *are* environments. . . . The principle of order, form, expressive energy, and finally of coherence for such environment-poems is the *phrase*, which I mean in a grammatical and in an extended gestural sense.
>
> *(103)*

The reader's puzzlement at this is alleviated a little when Fletcher explains that "environments are a special kind of natural ensemble, where drama and story are not the issue, where emotion is subordinate to the presentation of the aggregate relations of all participants, rather than the striking enhancement of singular or single heroes or heroines" (123). And then: "Waves of change will permit the poem and the nation to become a unity arising from a diversity, exactly as the aggregates of environment compose a composite unity" (143). Perceptive readers will note that Whitman's poems *not only* embody an idealized "environment," but *also* somehow epitomize "the nation" – the United States of America. Its democratic institutions, like Whitman's poems, are inherently ecological: "Nature's economy calls not for a House of Burgesses, of Lords, of Commons, but more radically for 'representatives' with whom we interact in a system of mutual co-representation " (124) Just as it's irrelevant whether or not Whitman *writes about the environment*, it doesn't matter whether America's political representatives actually enact any environmental legislation because the putative affinity is formal.

Of the many things one might say about this argument, here are just two: the House of Representatives, like all democratic bodies, "interacts" with the electorate that lends it legitimacy in specific, highly mediated ways. In twenty-first century America, these include intentional partisan gerrymandering. Furthermore, an "environment" is not a delineated or operative term in anyone's discourse, so to specify that it is a "natural ensemble" or that "emotion" (whose – the animals?) is subordinate to the presentation of the aggregate relations of all participants accomplishes nothing. The shift from mere "aboutness" to the apparently stronger claim that Whitman's poems "*are* environments" takes a weak analogy and turns it into a feeble metaphor, italics notwithstanding.

Exaggerated claims about the "ecological" qualities of one's favorite poet or species of poem is facilitated by the endlessly arguable quality of poetry itself, as we might expect, and that of ecology too, which we might not. Even as Fletcher was penning his improbable treatise, Phillips was warning that, "for ecocriticism to be of substance as an interdisciplinary field, it needs to realize that ecology is not a slush fund of fact, value and metaphor, but a less than fully coherent field with a very checkered past and a fairly uncertain future" (45). The accomplishment and originality of Whitman's poetic form is undermined, not elevated, by Fletcher's conflation of it with "environments" and American political institutions. Indeed, in 2020 it is hard to say which of these analogies is more ludicrous.

Motivated form, again

Taking a long view of the history of literary theory, it is striking that claims about the motivation of literary form repeat like Groundhog Day. "Free verse," or open form poetry, has been aligned repeatedly with various axes of human emancipation, while the changing form of the novel inspired decades of debate through the middle of the twentieth century about the ideological valence of realist and modernist alternatives. The official Soviet position advocated (and, within the Communist bloc, imposed) Socialist Realism, while Bertolt Brecht adapted for the purposes of class emancipation the dangerously apolitical concept of *ostranenie* or defamiliarization from Russian formalists such as Viktor Shklovsky. Estranging dramatic techniques, Brecht maintained, could disrupt and break the stranglehold of bourgeois ideologies on audience's minds, and were therefore to be approved rather than suppressed. Outside the Soviet bloc, where the question of motivated form seldom seemed a matter of life and death, the Frankfurt school took the side of modernist experimentation, which was seen as revealing the cracks and contradictions in the "totalizing" system of capitalism. Theodor Adorno, no ally of Brecht's, contrasts naturalistic absorption with modernist estrangement in his 1954 reflections on "The Position of the Narrator in the Contemporary Novel":

> The [realist narrator] was moral: taking a stand for or against characters in the novel. The new [modernist] reflection takes a stand against the lie of representation, actually against the narrator himself, who tries, as an extra-alert commentator on events, to correct his unavoidable way of proceeding. This destruction of form is inherent in the very meaning of form.
> *(Adorno and Nicholsen 34)*

The abyss of the Second World War, and the "administered world" that came after, made realist narrators impossible, even unethical. Far from having an obligation to represent class struggle realistically, art, for Adorno, ostentatiously withdraws from the structure of commodity production with which it is complicit. Its radicalism can only ever be formal, as Terry Eagleton explains:

> For art to *refer*, even protestingly, is for it to become instantly collusive with what it opposes. . . . Any positive enunciation is compromised by the very fact of being such; and it follows that what one is left with is the purest imprint of the gesture of negation itself, which must never stray from the elevated level of form to anything as lowly as content.
>
> (350)

The key point here is that, even though Adorno's esthetics is rigorously anti-instrumentalist (the work of art retains its frail, residual capacity to bring about change on the condition of its withdrawal from the arenas where it is most needed) it still imputes a motivation, albeit negative, to form per se. Reviewing this saga as a cautionary tale for ecocritics, Nancy Easterlin concludes that:

> the history of Marxist attempts to align ideological commitments with specific textual practices attests to the inadvisability of pinpointing a correct aesthetics, for the simple fact of the matter is that the semantic content of an artwork cannot be discerned on the basis of something so general as a style or a mode.
>
> (*Easterlin 99*)

One might add that contemporary scholars could benefit from a sense of history, or irony, when they claim that literary forms are perfectly fitted to, say, the climate crisis that have in the past been adduced in support of class struggle or women's liberation. Moreover, the position for ecocritics is conspicuously worse than it was for Marxists, who merely sought to explain how, and to what degree, the cultural "superstructure" was determined by the relations of production that constitute the socioeconomic "base" of society – a broad question for which motivated literary form is a paradigmatic case. Adding a world of wildly heterogeneous nonhuman actors into the "basic" relations of production would prove impossibly bewildering (Garrard "Literary Theory 101" 782). It is this conceptual canyon, between ecology and poetic form, which Fletcher blithely collapses when he asserts that Whitman's poem "*are* environments."

Close reading for estrangement

The dominant competing account of the value of ecopoetry derives ultimately from the early twentieth-century movement known as "Russian Formalism." Though the moniker came from the group's Soviet critics and was intended to be disparaging – what could be more bourgeois than abstraction and estheticism? – it accurately captured the group's interest in language, genre, and literary technique. Whereas everyday life tends to automatize both perception and language, literary language was defined, for Shklovksy, by its deployment of forms or "devices" that can startle us out of our stupor. As Shklovsky puts it, "[Art] exists that one may recover the sensation of life; it exists to make one feel things, to make the stone *stony*" (Lodge 20). Notice that this, too, makes a case for motivated form since Shklovsky also suggests "*Art is a way of experiencing the artfulness of an object; the object is not important*" (ibid). "Literariness" just names those devices that, at a given moment in history, can disrupt automatization; the content is not important.[1] However, defamiliarization has at least one signal advantage over the purveyors of "environment poems": poems are typically made of words, and have to be interpreted by human minds, so an account that draws on these characteristics is off to a good start.

John Felstiner's *Can Poetry Save the Earth? A Field Guide to Nature Poems* channels Shklovsky, albeit without attribution. Extolling the crystalline revelation of William Carlos Williams's "signature poem" ("So much depends/upon/a red wheel/barrow"), Felstiner avers:

> So much depends on seeing the things of our world afresh by saying them anew. Swamped by commerce and events – markets, movies, Internet, the world's confused alarms – we could do with poetry's exact enlivening touch for nature's common surprises.
>
> *(Felstiner 3)*

Absent the environmental moralism, there seems no reason poetry couldn't enliven our automatized perspectives on markets, movies, and the Internet too. Still, it is an appealing sentiment that "Poems make us stop, look, and listen long enough for imagination to act, connecting, committing ourselves to the only world we've got" (13). Scott Knickerbocker's *Ecopoetics: the language of nature, the nature of language* is, as its chiasmic subtitle indicates, less home-spun and more willing to accentuate rather than elide the "important gap between word and world" (Knickerbocker 9). Like Felstiner, though, Knickerbocker asserts that:

> Heightened perception promotes deep thinking. A literary critic's close reading, if it succeeds, is a form of deep thinking on paper about literary language. This deep thinking should welcome readers in and, like the poem itself, inspire, startle, or coax them into perceiving and considering anew the world as brought forth through poetry.
>
> *(18)*

Crucially, neither Felstiner nor Knickerbocker suggests that poetic form is, or could be, motivated independently of so-called content. As Derek Attridge puts it:

> Meaning is . . . not something that appears in defining opposition or complementary apposition to form, as it is conceived of in the aesthetic tradition, but as something already taken up within form; forms are made out of meanings quite as much as they are made out of sounds and shapes. Form and meaning both happen, and are part of the same happening.
>
> *(Attridge loc.2251)*

In the remainder of this essay, we examine two examples of ecopoetry in the light of the idea that form and meaning are "part of the same happening," and that this happening has the capacity to engender refreshed perception. In conclusion, we consider some limitations to the estrangement hypothesis and propose some practices of close reading that might alleviate them.

Returning birds, embracing Moon Jellies

Ecopoetry has a distinct academic reputation of gravitating toward sociopolitical meanings beyond the "words on the page." Its definition and scope are still very much in contention and have been evolving as the fields of ecocriticism and the environmental humanities have also broadened. In the "Editors' Preface" of *The Ecopoetry Anthology*, a substantial collection of American "ecopoetry," editors Ann Fisher-Wirth and Laura-Gray Street, acknowledge that the constant evolution of ecopoetry and how it has been defined are a process that should be seen "not as teleological but as a shifting landscape in which all forms of poetry have bearing and legitimacy, their presences and purposes ebbing and flowing in different communities and contexts" (Fisher-Wirth and Street xxix–xxx). Ecopoetry, then, is less often the victim of formalist "dummies" than instrumentalizing readings and writings that serve the eco-political needs of the present moment. As instructors around the world rewrite their curriculums to address environmental degradation and climate change, the dominant reading practice of ecopoems is not necessarily a "close" reading in Richards's sense of the word but rather a convenient identification of topical imagery. Poetry – especially figurative language – is treated as an infuriating puzzle or obstinate riddle to which the teacher holds the key. Rebelling against this view, the poet Don Paterson declares that, "we are born . . . into a condition of metaphor, a metaphor really being a contextual restriction of sense" (Paterson). Thus, reading poetry ought to return us to the primal scene of undifferentiated metaphoricity. Paterson further suggests that poets (just like any crafter or maker familiar with their materials) understand that "sound and sense are the same thing" and that "the acoustic and semantic properties of the word are not even interchangeable for [poets]; they are wholly consubstantial." In defiance of woolly subjectivism and instrumentalizing pedagogies, Paterson deploys a rhetoric of precision crafting: "a poem is just a little machine for remembering itself; a poem makes a fetish of its memorability" because every single word choice draws the reader to event, feeling, and image. With that in mind, the teaching of poetry ought to be a kind of reverse engineering, not emoting or assessing "relatability." Learning, and teaching, how "the ear thinks" is hard because it is simultaneously intuitive and objective. For instructors and students, it implies a necessary reconditioning.

With Paterson's strictures in mind, we turn to "The Birds Returned" (Holmes 215) by John Smith, a writer of the Canadian Maritime provinces. It is a seemingly nonmetrical ecopoem that narrates the changes in a "way station" island that used to be a breeding ground for an unspecified type of bird. On an initial reading, the poem can easily be interpreted in familiar elegiac terms (Heise): the changing landscape of that island, brought about by men, has repelled these birds and they are now, unfortunately, a distant memory. Close attention to the melding of sound and sense in this poem, though, suggests a more ironic, even sardonic, reading.

The first stanza seems to narrate, in abstract terms, the evolutionary process of adaptive radiation. Here it is with stressed syllables bolded:

> The **birds** re**turned** so *often to the* **is**land, to **one**
> par**ti**cular **tract** of ***peat*** *bog, that they* **lost** the **hang** of
> inter**breed**ing with **an**y **oth**er **race** than **those**
> who **felt** at **home** with the ***sa****vour of a* **spe**cial **ground**-spruce.
> *(1–4)*

The strung-out five stress meter could be read as having lots of *really* long measures (quadruplets italicized) or, on a more isochronous reading, as veering often to six stresses per line. Given the subliminal "rightness" of four- and five-stress duplet meters to experienced readers of English poetry, this will be sensed as either too many stresses or overly elongated measures. Those moments, with their puzzling tension, combine to impute a certain bizarre perversity to the birds: they keep coming back until they are too specialized to survive anywhere else. Reproductive isolation is represented as losing "the hang/of interbreeding" and a unique affinity for a specific "savor." This "special ground-spruce" recalls "peat-bog" both rhythmically and in its evocation of a certain obsessive particularity.

The next stanza begins with a striking enjambment that shifts the sense of the preceding sentence. The olfactory and gustatory imagery of the weather-beaten spruce is suddenly reframed by the tactile sense "that a man . . . could walk on top of it and not fall through." Even as that's going on, the interjection "– but there were no men –" stresses that this is an experience that no one, in fact, could have. Until, "Inevitably," they do. Again, a wobbly five stress meter prevails, with a rhythmic and rhetorical parallel between "were no men" and "not fall through" – the repeated litotes nests a negation within a negation, while both can be read as duplet measures only by demoting the middle syllable. Here we see, as Paterson suggests, that "Poetry is the paradox of language turned against its own declared purpose, that of nailing down the human dream."

As we'd expect of an environmental elegy, "Men" disrupt the "particular," "special" reciprocity of birds and place: "Then everything changed –" Again, there is a reframing enjambment, but this time it is a further litotes, a telling qualification: "**not** by **much**, but e**nough** to keep **wa**riness/on **wing**, **fly**ing **wi**der **cir**cles, **land**ing **rare**ly . . . " The lines shorten, seeming to fall short of five stresses rather than etiolating the measures, landing hard on the singlet "wing" before the forlorn repetition of five duplets signal the birds' spiral to oblivion. The assonance of "much" and "enough" underlines the subtle devastation of anthropogenic impacts such as trophic cascades, while the alliterative "w" – continued, ironically from the human "way station" of the previous stanza – connects fear and flight. Then, too, "flying" is linked by alliteration to "farther," "wider" by assonance to "desired," to drive home the deadly tension of need and aversion.

The second tercet has another surprise in store: a shift into metanarrative: "These are depictions of perhaps the last/breeding pairs." What are? These "images" evoked, ekphrastically, by the words of this poem, or some other representation that they, in turn, convey? The extinguished birds recede into a romanticized past: "**Stor**ies/**tell** of a **land** of **one** pro**lif**ic **moun**tain, **back** // be***fore*** ***diff****erence, be***fore** things **part**ed from their **names**." Rhythmically, the italicized singlet/triplet combination highlights the irony of narrating what came *before*, in Paterson's words, "the fall into language, asymmetry, the observation that we are other than the breast, the mother and the back garden." The verse structure of the poem, too, emphasizes mournful separation: two quatrains and two tercets make a sonnet, of course, but then there's that extra line on its own, seeming to parallel the surplus stresses and syllables throughout the poem. It is the old story of language supposedly dividing us from Nature, which Robert Watson traces back to the Renaissance (Watson). Or is it? Just as the poem seems to hold the birds responsible for their vulnerability to extinction, it is the things that leave their names behind, not the other way around. There is a rhythmic ruction there too: one wants to stress "things," to emphasize the contrast with names, but three stresses in a row is a train wreck in a five-stress line. It's followed by a quadruplet measure that begs for "from" to be promoted to restore the duplet meter. All this to say, by every means available to poetry: why did you choose frailty, birds? Why did things abandon their names? These are not rhetorical questions.

A poem that seems to embody Romantic nature elegy turns out, on close reading, to ironize and estrange it. Clearly, though, a quick or instrumentalizing approach ("I just need some quotes for my essay . . . ") will be biased toward the first, not the second, of these, while advocacy for motivated form will presumably discard any poem still in thrall, however restlessly, to iambic pentameter. The interdependence of form and meaning enacted here requires a slow reading that is oftentimes forgotten in the time-limited, distracted environment of the twenty-first-century classroom: the patience and preparedness to hear before the poem can speak, and the tenacity to teach close reading skills carefully.

A more playful offering, Washington State writer Nancy Pagh's "Moon Jelly" (Blomer), is an open form, or free verse, constructed of five imagistic quatrains. It captures the sensory experience of diving and swimming with moon jellies, or the *Aurelia aurita*, a jellyfish that has four gonads visible through its translucent, bell-shaped body. Despite being non-metrical, it begins with a four-stress duplet line:

> No **stars** or **moon** in **soot**y **sky**;
> **rain falls** through **ced**ars
> in the **mouth**
> of **Jors**ted **Creek**.
>
> *(1–4)*

Arriving at three stresses in succession across the first line break, the brief forward momentum is arrested at the aural and visual image of falling rain (this is the Pacific Northwest!), only to accelerate through a quadruplet to land on the word concluding the shortest line: "mouth." Already, the experienced reader might suspect a poetic resurrection to uncanny literality of this dead metaphor for a watercourse's opening.

The next couple of stanzas build through visual images, which are rendered lightly ironic by the ambient darkness and the jellies, "transparent/made of tissue, or fog/frail as soap bubbles." Alerted by "mouth," we notice "the open/hungry anemones," as the night swimmer is "embraced by moon jellies," an image of tactile reciprocity and belonging. All of a sudden, the lyric voice seems to address the creatures:

> Oh, **moon jell**y;
> a **name** to **spread** on **ev**ening **toast**
> and **eat**
> **bite** by **ti**ny **bite**.
>
> *(17–20)*

The apostrophe is directed at the *name*, though, not the creatures themselves, in a gesture of enthusiasm and desire that borders on erotic. Following the brief, clotted first line, the rhythm pours through four duplets onto the arresting line-ending singlet "eat." Thanks to the way the creatures' name conjoins the unearthly and the everyday, the desire to consume it, delicately but with relish, seems to ally the speaker with all the other hungry mouths in the poem. Language, in this context, binds us to the world rather than estranging us, in something like the way Gary Snyder extols "Language Goes Two Ways." While he acknowledges the power of language to dominate and alienate, Snyder also points out that "Languages were not the intellectual inventions of archaic schoolteachers, but are naturally evolved wild systems whose complexity eludes the descriptive attempts of the rational mind" (Coupe 127). "Moon Jelly" reaffirms one of the core issues in ecopoetry: the constant attempt to "get into the right relationship" with the cycles of lives and their renewal, which Robert Hass suggests is the oldest impulse of poetic language (Fisher-Wirth and Street lx). As we have shown, form and language together *constitute* the meaning, which could be summed as the experience of both naming and eating as forms of interconnection. It is striking that two such different poems as "The Birds Returning" and "Moon Jelly" – the one abstract, loosely formal, almost allegorical; the other commingling the numinous and the visceral – come back at their conclusions to reflexive contemplation of the material of poetry: language.

Close reading as eco-pedagogy

Our critique of the myth of motivated form implies that there is no such thing as "ecological form," as such. It makes us suspicious of claims such as Matthew Griffiths's in *The New Poetics of Climate Change*: "Just as climate change disrupts and disturbs pastoral or Romantic conceptions of nature, so formally experimental, avant-garde writing in the Modernist tradition can offer the opportunity to rethink rather than reinforce reader-writer relations" (Griffiths 5). As we have seen, this is just the latest in a long line of apologies for (and critiques of) avant-garde poetics. Even if true, it would not follow that, absent some interest in or knowledge of earth systems science, writers adopting modernist forms are uniquely well-suited to the immense challenge of representing climate change. Griffiths's final chapter surveying contemporary "climate change poetry" tacitly acknowledges that topicality matters, although it continues to privilege "innovative" and "sophisticated" (160) open form over the putative constraints of the lyric tradition. Timothy Clark, who is well-disposed toward Griffiths's project, admits that the "obscurity" typical of avant-garde poetry "must be worrying for any ambition that ecopoetry can be a significant kind of environmental activism, given that even classes on traditional poetry can already too easily become a kind of crossword puzzle-solving exercise" (Clark 59).

In contemporary Anglophone schools and universities, poetry is at risk both from this debased version of Richards's practical criticism and, less obviously, from its instrumentalization in theory-led and identity-centered criticism. As Derek Attridge acknowledges in *The Singularity of Literature*, the resistance to instrumentalism expected of literary studies is counterbalanced – sometimes overwhelmed – by the exigencies of the academic job market:

> Graduate students frequently choose courses or dissertation topics, with eminent good sense, on the basis of their usefulness for their chosen career, rather than on the intellectual provocations and rewards they might offer; articles and books are written with an eye to the marketplace and the syllabus; and "theoretical approaches" are mastered (or their salient catch-phrases learned) in order to utilize them efficiently in reading and writing instead of being approached with an openness that allows for a range of possible outcomes – including a challenge to the very project they are supposed to be serving.
>
> *(loc.349)*

However hard it may be to read Deleuze and Guattari, close reading poetry is harder because it is simultaneously cerebral and instinctual. Scansion especially, having been made artificially challenging by irrelevant Greek terminology, is unfamiliar even to graduate students of English. Though poems may be attractively brief, attentive responses to them cannot be hurried. Far quicker and easier to pick out quotations to illustrate theses derived from other, somehow more authoritative, texts. (The theory goes in the introduction, like this, with the poetry arriving later to confirm its rightness. Poetry is meek and compliant; it seldom answers back.) In this context, it is all too predictable that a STEM-centered version of eco-literacy will reprise a limited role for poetry, as for example David Orr does in drawing from literature (*Moby-Dick, Frankenstein*, Wendell Berry) for moralistic anecdotes and proverbs but excluding it from the core curriculum of environmental education. While he allows that students should be encouraged to read as well as growing food and learning to use solar energy, Orr defines a canon intended to function didactically, not critically (Orr).

By contrast, we argue that close reading of poetry deserves a place at the center of our conception of eco-literacy, not the periphery. With that affirmation, we not only assert the importance of beauty, value, history, language, and form to environmentalist struggles, we also reject Orr's dire, reductive prescription for eco-literacy that has no place for students as active learners – "prosumers" of knowledge and wisdom, as the current jargon has it. We seek to redefine the relationship of education and activism in line with Attridge's brilliant synthesis of Russian Formalism, liberal humanism, and poststructuralist theory. Above all, *The Singularity of Literature* seeks to understand the *meaning* (not the psychology) of creativity, as Attridge highlights in this parenthetical example:

> (One of Coetzee's characters reflects with wonder on Shakespeare's creation of a particular phrase: "Out of the dark emerging, out of nowhere: first not there, then there, like a newborn child, heart working, brain working, all the processes of that intricate electrochemical labyrinth working. A miracle.")
>
> *(loc.516)*

Acknowledging the historical relativity of art's *ostranenie*, he summarizes that "[t]o create an artwork, then, is to bring into existence a configuration of cultural materials that, at least to a certain group and for a certain time, holds out the possibility of a repeated encounter with alterity" (loc.701). Singularity is Attridge's name for the quality that enables emergence, which, as we have seen, conjoins form and meaning as "one happening" and which makes possible, for a select few artworks, a seemingly perpetual *reemergence* into significant otherness.

The eco-pedagogical counterpart to Attridge's literary theory is what Jonathan Skinner, editor of the journal *Ecopoetics*, calls "a practice of emergency." The phrase encompasses both the familiar sense of environmental crisis and the prospect of "emergence of new forms of life" (Hume 756). Drawing on Skinner, Garrard outlines an expansive "unprecedented ecopedagogy" that "conjoin[s] environmentalist objectives with progressive humanistic educational techniques and ideals" (Garrard "Towards an Unprecedented Ecocritical Pedagogy" 199). Teaching poetic form in this context demands reading practices that are not only close but, crucially, *slow*. As Richard Kerridge puts it, "A text is for life, not just for the degree. Here is 'slow reading' to go with 'slow food'" (Kerridge 21). Close reading is also, as Scott Knickerbocker observes, an embodied practice, to which teachers should draw attention:

> When a poet skillfully experiments with various sound effects in a poem, when a reader revels in the sensuous pleasure a poem provides, and when

> a literary critic deepens our understanding or appreciation of the way a poem's form – its body – shapes meaning, these writers and readers experience both their own and the poem's embodiment, even when silently sounding the poem to one's inner ear.
>
> *(Knickerbocker 7)*

We have come full circle back to IA Richards's concern with failures of "sensuous apprehension."

In all these ways, an ecological literacy informed by practices of close reading is not a rushed methodology to find solutions and act on them, nor sign up to a theory and find quotations to support it; rather, it fosters a return to broad, deep literacy within poetry and across disciplines, and an acknowledgment of the irreducible tensions between the art form and its environment. It centers poetic form, not as a motivated phenomenon operating independently from meaning, but as a discipline of intelligence and intuition that reveals language as at once wild and strange.

Note

1 Shklovsky's conception of estrangement is, of course, historically and culturally relative, as Terry Eagleton points out: "One person's norm may be another's deviation: "ginnel" for "alleyway" may be poetic in Brighton but ordinary language in Barnsley" (*Literary Theory: An Introduction*, Oxford: Blackwell, 1983, p. 5).

Works cited

Adorno, Theodor W., and Shierry Weber Nicholsen. *Notes to Literature*. Ed. Tiedemann, Rolf. New York: Columbia University Press, 2019. Print.

Attridge, Derek. *The Singularity of Literature*. London, United Kingdom: Psychology Press, 2004. Print.

Blomer, Yvonne, ed. *Refugium: Poems for the Pacific*. Halfmoon Bay, BC: Caitlin Press, 2017. Print.

Clark, Timothy. *The Value of Ecocriticism*. London: Bloomsbury, 2019. Print.

Coupe, Laurence. *The Green Studies Reader: From Romanticism to Ecocriticism*. London: Routledge, 2000. Print.

Eagleton, Terry. *The Ideology of the Aesthetic*. Oxford: Blackwell, 1995. Print.

Easterlin, Nancy. *A Biocultural Approach to Literary Theory and Interpretation*. Baltimore, MD: Johns Hopkins University Press, 2012. Print.

Felstiner, John. *Can Poetry Save the Earth?: A Field Guide to Nature Poems*. New Haven: Yale University Press, 2009. Print.

Fisher-Wirth, Ann W., and Laura-Gray Street. *The Ecopoetry Anthology*. San Antonio, TX: Trinity University Press, 2013. Print.

Fletcher, Angus. *A New Theory for American Poetry: Democracy, the Environment, and the Future of Imagination*. Cambridge, MA; London: Harvard University Press, 2004. Print.

Garrard, Greg. "Literary Theory 101." *Interdisciplinary Studies in Literature and Environment* 17.4 (2010): 780–83. Print.

———. "Towards an Unprecedented Ecocritical Pedagogy." *Teaching the New English*. Ed. Knights, Ben. London: Palgrave Macmillan, 2017. Print.

Glotfelty, Cheryll, and Harold Fromm. *The Ecocriticism Reader: Landmarks in Literary Ecology*. Athens; London: University of Georgia Press, 1996. Print.

Griffiths, Matthew. *The New Poetics of Climate Change: Modernist Aesthetics for a Warming World*. London: Bloomsbury, 2017. Print.

Heise, Ursula K. *Imagining Extinction: The Cultural Meanings of Endangered Species*. Chicago: The University of Chicago Press, 2016. Print.

Holmes, Nancy, ed. *Open Wide a Wilderness: Canadian Nature Poems*. Waterloo, ON: Wilfred Laurier Press, 2009. Print.

Hume, Angela. "Imagining Ecopoetics: An Interview with Robert Hass, Brenda Hillman, Evelyn Reilly, and Jonathan Skinner." *Interdisciplinary Studies in Literature and Environment* 19.4 (2012): 751–66. Print.

Kerridge, Richard. "Ecocriticism and the Mission of 'English'." *Teaching Ecocriticism and Green Cultural Studies*. Ed. Garrard, Greg. Basingstoke, Hants.: Palgrave Macmillan, 2012. Print.

Knickerbocker, Scott. *Ecopoetics: The Language of Nature, the Nature of Language*. Amherst: University of Massachusetts Press, 2012. Print.

Kricher, John C. *The Balance of Nature: Ecology's Enduring Myth*. 2009. Web.

Lodge, David. *Modern Criticism and Theory: A Reader*. London: Longman, 1988. Print.

Orr, David W. *Ecological Literacy: Education and the Transition to a Postmodern World*. Suny Series in Constructive Postmodern Thought. Albany: State University of New York Press, 1992. Print.

Paterson, Don. "Rhyme and Reason." *The Guardian*. 6 November 2004. Web. 12 September 2020.

Phillips, Dana. *The Truth of Ecology: Nature, Culture, and Literature in America*. New York; Oxford: Oxford University Press, 2003. Print.

Richards, I. A. *Practical Criticism; A Study of Literary Judgment*. London: Kegan Paul, Trench, Trubner & Co., 1929. Print.

Watson, Robert N. *Back to Nature: The Green and the Real in the Late Renaissance*. Philadelphia, PA: University of Pennsylvania Press, 2006. Print.

3

ASSEMBLING THE ARCHIVE

Close(ly) reading great auk extinction with Walton Ford

Nicole M. Merola

> *as if God had made the innocencie of so poor a creature to become such an admirable instrument for the sustenation of man.*
>
> – Captain Richard Whitbourne, *A discourse and discovery of New-found-land . . .*

Pinguinus impennis (great auk) went extinct in the mid-nineteenth century, a casualty of cumulative consequences of evolution, transatlantic exploration and imperial expansion, and patterns of overconsumption by certain groups of humans in industrializing societies. The *IUCN Red List of Threatened Species* marks 1844 as the death date of the last known pair, which was killed during a raiding party commissioned by collector Herr Carl Siemsen (Fuller 80), and 1852 as the sighting of the last live great auk (Birdlife). A flightless North Atlantic seabird that inhabited low-Arctic and boreal regions and came ashore in large colonies only to incubate one egg per year per breeding pair, the great auk's misfortune was to congregate and reproduce in locations relatively accessible to human predators. Evidence from prehistoric middens indicates use of the great auk by aboriginal peoples in northern Europe and Newfoundland. Evidence from transatlantic voyages in the sixteenth, seventeenth, and eighteenth centuries; interviews in the nineteenth century with people engaged in great auk harvesting; studies of museum specimens; and consideration of the habitat ranges of other North Atlantic seabirds locates former breeding colony sites across the northeastern and northwestern Atlantic, including on islands off the coast of Greenland, Iceland, Ireland, and Scotland; on the Magdalen Islands in the Gulf of St. Lawrence; and on Funk Island, off the eastern coast of Newfoundland, among other places. The Funk Island colony was the largest, with an estimated population of 200,000 great auks.[1]

The *IUCN Red List's* hybrid date 1844/1852 marks an irrevocable rupture in evolution, the end point of the multi-temporal process of extinction of *P. impennis* (Birdlife). To specify: a 300–400-year period of human–great auk relationship centered on anthropogenic violence – harvesting, death, and consumption of nonhuman animal life – scythed evolutionary time. Three specific forms of violence drove the great auk to extinction.[2] The first centered on uses of the great auk for sustenance: eating bird eggs, eating bird meat, using bird meat as fish bait, and rendering birds for oil. The second involved harvesting feathers for use in down pillows and featherbeds. The third encompassed specimen collecting for display by museums and individuals.[3] These three forms existed simultaneously and sequentially as an amplifying feedback loop. For instance, feather harvesting and eating birds would have been simultaneous activities; feathers would have been plucked in order to prepare the birds as food. But feather harvesting also occurred without further use of the bird, as indicated by eighteenth-century sailor Aaron Thomas: "If you come for their Feathers you do not give yourself the trouble of killing them, but lay hold of one and pluck the best of the Feathers. You then turn the poor Penguin adrift, with his skin half naked and torn off, to perish at his leisure."[4] As an example of sequentiality and a familiar tale of scarcity ratcheting up demand and price, the practice of specimen collection accelerated toward the end of the extinction process.[5] Although their aims might have been slightly different – metabolism, comfort, display – these harvesting activities were all underwritten by attitudes like those Captain Richard Whitbourne evinces in this chapter's epigraph: the belief that the "sustenation of man" renders all other creatures into standing reserve.[6]

Extinction studies are flourishing in the environmental humanities. This work resoundingly rebuts instrumentalist, speciesist attitudes like those that caused great auk extinction and, instead, explicitly values life forms and ecosystems from across the micro- and macro-biological spectrum. Recent monographs, anthologies, journal issues, and articles in this vein include an *Australian Humanities Review* special issue (Rose and van Dooren 2011), *Wild Dog Dreaming: Love and Extinction* (Rose 2011), "Aesthetics of Loss: Biodiversity, Banal Violence and Biotic Subjects" (Yusoff 2012), *The Anthropology of Extinction: Essays on Culture and Species Death* (Sodikoff 2012), *Flight Ways: Life and Loss at the Edge of Extinction* (van Dooren 2014), *Essays on Extinction* (vols. 1 and 2, Colebrook 2014), *Imagining Extinction: The Cultural Meanings of Endangered Species* (Heise 2016), "Each Time Unique: The Poetics of Extinction" (Wolfe 2017), "Photographing the Last Animal" (Schuster 2017), *Extinction Studies: Stories of Time, Death and Generations* (Rose, van Dooren, and Chrulew 2017), and *After Extinction* (Grusin 2018). This work, together with contemporary popular nonfiction focused on extinction (e.g., Quammen 1996, Ellis 2004, Weisman 2007, Kolbert 2014, Brannen 2017), attests to widespread interest in the subject. Although each works in a slightly different register, extinction studies texts generally explore two related sets of questions. First, they ask how we identify, define, and map different kinds of extinction. Most texts explicitly or implicitly consider some version of the categorical question what kind of an event is extinction, which bundles existential concerns with concerns about data, interpretation, representation, and communication. Second, they examine how cultural and scientific texts craft various responses to ongoing biodiversity losses. To offer a partial list, these might include assessments of ecosystem value, arguments for new conservation practices, arguments for new legislation, experiments in speculative philosophy, and the creation of fine arts objects, my subject below. One critical task of extinction studies is to tack between contours common across different extinction events and those that are singular. The kinds of story templates Ursula Heise tracks and the tropes Joshua Schuster delineates are examples of the former. Concepts like "forms of life" (van Dooren) or "each time unique" (Wolfe, following Derrida) offer tools for mapping the latter, even as these concepts become fixtures in extinction studies. Furthermore, given the anthropogenic origins, scope, and speed of the sixth extinction, extinction studies also often enfolds discussions of environmental affect.

Since the early 1990s, contemporary U.S. artist Walton Ford has been making exquisite, detailed, narrative paintings and prints that engage and critique conventions of natural history illustration through staging violent scenes of encounter between animals. More specifically, Ford works in direct conversation with masculinist, romantic aspects of eighteenth- and nineteenth-century exploration and natural history (e.g., mythologies of the explorer, practices of specimen collection, attitudes of human exceptionalism), with the work of John James Audubon, with extinction, and with colonialism. He also employs as visual subjects specific historical moments or anecdotes of cross-species encounter. These categories of attention often overlap in a single piece. While some of Ford's work maps intraspecies violence, his predominant concern is cross-species conflict. In some cases, nonhuman animals are the only participants in the frame; in others, human animals are directly or indirectly implicated.

A number of Ford's paintings and prints directly cite, comment on, and extend illustrations from Audubon's *The Birds of America.* Of these, Ford's Carolina parakeet, flamingo, great auk, passenger pigeon, and turkey images are especially notable.[7] His two great auk paintings – "Funk Island" (1998) and "The Witch of St. Kilda" (2005) – tell related but different extinction stories. "Funk Island" collects a set of repeated actions of great auk slaughter into one dramatic image that distills the destruction of the island's great auk colony; the anthology's cover includes a detail from "Funk Island." "The Witch of St. Kilda" draws from a well-known anecdote to stage a moment in one bird's life. Both paintings allegorize the extinction of the entire species. In addition to the concerns detailed above, "Funk Island" and "The Witch" demonstrate Ford's interest in anecdote, archival research (material and textual), citation, dark humor, history, intertextuality, narrative, and palimpsest.

While Ford's interest in anthropogenic extinction has developed concurrently with the concept of the Anthropocene in its contemporary form, his exhibitions are regularly reviewed, and he is frequently interviewed, scholarly consideration of his work has been limited.[8] So, one aim here is to illustrate Ford's importance for Anthropocene conversations, in general, and extinction studies, in particular. More precisely, I argue that Ford's extinction-focused objects rehearse located and historicized Anthropocene stories; to use Elizabeth DeLoughrey's formulation, he provincializes the Anthropocene.[9] In this sense, his work not only brings a concrete and distinctive Anthropocene into view, but "Funk Island" and "The Witch," in particular, also serve as examples for modes of eco-studies reading in/of/for the Anthropocene. Since its emergence, the idea of the Anthropocene has been sutured to questions of inscription and reading. Semiotic metaphors populate Anthropocene discourse across the natural sciences, social sciences, and humanities. For instance, designating the Anthropocene requires discerning what anthropogenic signals are different enough from background earth system processes to denote geologically significant difference within the stratigraphic record. In other words, geologists read earth texts to look for durable anthropogenic marks on the earth's material surfaces and in its biogeochemical processes. During the relatively climatologically stable Holocene, which enabled settled societies and population growth, a sequential temporality of writing and reading earth systems prevailed. To shorthand, the planet did something (suddenly or gradually) and humans, other animals, and plants responded (quickly or slowly). The Anthropocene jumbles these sequential practices of reading earth systems such that reading, writing, rereading, writing speculatively (future scenario modeling), and debating forms of overwriting earth system processes (geoengineering) occur and recur simultaneously. Time, or rather, sequential earth–human relations, are out of joint.[10] This simultaneity can be anxiogenic, but it also reveals an important point the Holocene's relative stability masked: the earth's fundamental dynamism.[11]

The "close(ly)" in my chapter title signals multiply, indexing that I value extended attention to Ford's work. I position Ford – an avid researcher, reader of written texts and animal bodies, and deployer of self-consciously literary gestures – as an artist who carefully parses extinction stories. And finally, eschewing the more familiar "close," I evoke investment in ongoing debates among literary critics about "proper" reading methods for approaching our objects: close, contextualized, descriptive, distant, formalist/new formalist, historicized, ideological, reparative, surface, symptomatic, and so on. Like analogous debates about the value of the humanities, these recycle. In eco-literary studies, conversations about how to apprehend our objects are additionally inflected by existential questions about the value of reading in the midst of socio-ecological calamity. None of these discussions are resolvable.

In addition to these concerns, I am interested in how well interpretive practices from literary studies transfer to other media and/or to other disciplines. Since many of the texts necessary for conceptualizing and pressuring Anthropocene stories are transdisciplinary and nonliterary (e.g., stratigraphic layers, ice cores, land use changes, sea-level rise, greenhouse gas accumulations, climate change models, and other visualized forms of scientific data), attending to the contours of reading practices across disciplines is especially pressing. Two questions for consideration are: how do form, medium, content, and scale construct persuasive and affective claims? How do we describe, map, illuminate, and contextualize Anthropocene affects, data, discourse, effects, and narratives in and via cultural objects?[12]

What I do below – practice reading in the Anthropocene as an exercise in valuing sustained attention to two great auk visual texts – is modest and exploratory.[13] I see reading and writing as always provisional, iterative practices that should generate dialogue. I think carefully discerning what a text offers (not what one might want it to offer) and parsing how it does so is intrinsically valuable. I employ lengthy description as a tool for noticing detail in Ford's paintings and a necessity for translating visual images into language.[14] And as I examine and interpret Ford's paintings, Carolyn Lesjak's description of critical practice, part of a trenchant critique of surface reading, is animating:

> what is needed is a better way of reading surfaces as perverse rather than as obvious, as never identical to themselves in their "thereness," and always found within and constitutive of complex spatial relations, both seen and not seen, deep and lateral, material and figural – all of which requires a more rather than less expansive reading practice: more interpretation, more dialectical complexity, a more rather than less invested critical position, because *relations*, after all, cannot be seen in any solely literal sense.
>
> *(251, emphasis in original)*

She values "putting [texts] in motion in three-dimensional space" (251). I think Ford's labor and artworks materialize the kind of practice Lesjak details. More specifically, I suggest his work puts texts in conversation across media and historical periods in the service of assembling an extinction archive that lingers horrifically.

"Funk Island" is a large-scale painting (roughly 5 feet tall by 10 feet wide) that combines watercolor, gouache, pencil, and ink on paper. In the upper left corner Ford uses cursive reminiscent of nineteenth-century writing to present the enhanced title: "Funk Island ~ or Good Conduct Well Chastised (1791)," a reference to Marquis de Sade's novel *Justine, or Good Conduct Well Chastised*. The painting's protagonists are members of Funk Island's great auk colony; they occupy the entire foreground. The great auks stand on a low large rock, marked by guano, that inclines from the lower left corner of the painting, flattens a bit in the middle with the surface tipped toward the viewer, and descends to the lower right corner. Their line of movement follows the rock surface, left to right, up, over, and down. Their destination is a fire generated by an unseen source located beyond the frame's right side. Three great auks are silhouetted as they walk into the fire. In the far background in the lower left corner a pair of ships and a small rowboat float on a placid ocean at the horizon line. One thin, gray, horizontal cloud smudges the air over the boats.

A plume of smoke from the anthropogenic fire undulates across the frame, starting behind the great auk colony and then rising to the upper left corner of the painting to undercurl the title. In the middle ground and background on the painting's right side, rocky cliffs rise. Northern gannets (*Morus bassanus*) perch on ledges across these guano-stained cliffs. They also soar above, in front of, and beneath the smoke in the middle ground and, in the far background, over the ships. Their ability to fly enables their status as spectators, rather than victims, of the foreground activity. In the smoke plume, Ford tucks gestural drawings of human pairs copulating. The visual style he employs for the human bodies – spectral, billowy, impressionistic outlines that share the colors and line weights of the smoke plume – makes these figures easy to miss. In contrast, the bodies of the great auks and northern gannet and the rocky outcrop and cliffs of their habitat are rendered with great detail to avian phenotype and geological texture.

Ford includes 42 great auks in "Funk Island." Reader, I counted them. Three times. In terms of their bodily orientation, the birds divide into three groups. With the exception of one great auk that directly faces the viewer, those in the left third of the frame generally face the backs of the birds in front of them. Those in the middle face a variety of directions – a few show their backs to the viewer, a pair looks directly at each other with their beaks open, in the nearest foreground one bird lies prone and another looks at it, one looks directly at the viewer with its wings spread, one looks at the sky with its wings spread, and two look at the right side of the frame. The great auks on the painting's right side are turned toward the fire; one of these raises its open beak to the sky.

Beneath the visual narrative, Ford includes a strip of handwritten text in the same style as the title. Three great auk names at its center are clear: "Garefowl or Great auk ~ Alca impennis."[15] Through their inclusion this strip offers an additional subtitle and echoes how eighteenth- and nineteenth-century natural history illustrations were often labeled. It also functions as a location for Ford to allude to a referent for the painting while grotesquely troping the field note; sentences from Aaron Thomas's 1794 *History of Newfoundland*, partially quoted earlier, surround the great auk's names. In full, the passage reads: "If you come for their Feathers you do not give yourself the trouble of killing them, but lay hold of one and pluck the best of the Feathers. You then turn the poor Penguin adrift, with his skin half naked and torn off, to perish at his leisure. . . . While you abide on this island you are in the constant practize of horrid cruelties for you not only Skin them Alive, but you burn them Alive also to cook their Bodies with. You take a kettle with you into which you put a Penguin or two, you kindle a fire under it, and this fire is absolutely made of the unfortunate Pengujins themselves. Their bodys being oily soon produce a Flame; there is no wood on the island." Ford uses techniques of fake-weathering at the painting's top edge, in spots of faux-foxing in the sky, and in the caption strip. As a result, parts of the Thomas quotation are faded.[16]

In "Funk Island" Ford consolidates into one terrible image of great auk destruction the repeated patterns of overharvesting that decimated the island's colony. "The Witch of St. Kilda" (roughly 1½ feet tall by 2½ feet wide, watercolor, gouache, pencil, and ink on paper) portrays a smaller, but still devastating tale. The painting's handwritten title – "The Witch of St. Kilda ~ 1840" – refers to a discrete incident, included in *A Vertebrate Fauna of the Outer Hebrides* (Harvie-Brown and Buckley), and regularly repeated in great auk lore. Briefly, in July five St. Kildan men found a great auk asleep on a ledge on Stac-an-Armin, a sea stack in the St. Kilda archipelago regularly visited to harvest great auks, northern gannets, puffins, and each species' eggs. They caught it, tied its legs together, and kept it captive in a bothy, where the great auk made "a great noise," would open its beak when humans approached, and tried to use its bill to sever the men's rope. During its confinement "a storm arose, and that, together with the size of the bird and the noise it made, caused [the men] to think it was a witch." The men beat the bird to death, killing it three days after its capture, and discarded its corpse behind the bothy.[17]

"The Witch" isolates one post-capture moment from this multiday event. Its eponymous great auk lies on its side on a flat rock that declines toward the bottom of the frame. Together the rock and the great auk occupy half the frame. The bird's legs are tied together with a rope, which stretches taut, casts a visible shadow on the rock, and runs out of the right side of the frame; its attachment point is unknown. Its tautness suggests the auk will be pulled out of the frame at any moment. Both of the great auk's wings extend forward toward the viewer. Its bottom wing rests on the rock while the top is in the air. Its upper chest and neck are bent; the crown of its head reaches toward the top edge of the painting. Its head turns in three-quarter profile. Its beak is slightly open with its tongue visible. With its one visible eye the great auk looks directly at the viewer. In both "The Witch" and "Funk Island" Ford precisely details great auk bodies, in particular their plumage (feather color, position, and texture) and beak, wing, and foot shape. The single body in "The Witch" affords Ford the chance for augmented attention to the grooves of the great auk's beak and the skeletal structure of and texture of the skin on its feet.

As in "Funk Island," Ford carefully attends to biological, geological, and hydrological elements of the great auk's habitat. The rock underneath the great auk is devoid of vegetation and its texture echoes the ocean's surface. The guano-stained cliffs in the middle ground on the right side offer ledges for four northern gannets to perch; three are aloft in a mostly featureless sky. The end of the cliff in the background displays a distinctive form of weathering. Toward its end it rises to a double-apex and then descends ski-jump-like to another point. The body of the auk blocks the rest of the cliff's descent into the ocean. The geological elements in the painting are indifferent to the life in the habitat they bedrock, referencing only deep time and non-anthropogenic creative and destructive forces. On the other hand, Ford takes advantage of the mercurial temporalities of weather to suggest the great auk's impending demise. From the left corner foreground, where the ocean meets the rock's edge, to the horizon, white-capped waves cover its surface. In the foreground it hues yellowish-green, darkening to turquoise, and then abyssal navy blue at the horizon. Poised to overtake the yellow glow of sunset to its immediate right, a storm cloud with rain streaks descending perches at the frame's left edge. The caption strip at the bottom of "The Witch" differs from the one in "Funk Island" in two important ways. First, to signal a 1791 reclassification, *P. impennis* has replaced *Alca impennis*. Second, Ford includes no faux field notes. Rather, the title, rope around the bird's legs, tumultuous ocean, and storm do all the allusive work to direct the reader to the source material.

Careful description of "Funk Island" and "The Witch" excavates Ford's diligence in rendering great auk habitats and their fragile and vulnerable bodies in order to position them as grievable subjects. Closely reading also reveals his allusive tendencies, which pull anecdote, archive, citation, geographical location, history, and humans into the frame while also linking the images to wider contexts. For instance, both painting titles name places with the dubious distinction of having hosted an identifiable event (of long or short duration) that clearly contributed to great auk extinction. While Ford includes no human figures in either painting's frame, the boats and fire in "Funk Island" and the rope in "The Witch" clearly signal their immediate proximity. These gestures make it impossible to disavow the anthropogenic origins of the violence the paintings depict. Together with the paintings' titles and the content they place squarely within the frame, these indicants suggest the full story exceeds the accounting capacities of these paintings, of painting in general, and more broadly, of any media or cultural object. In other words, unlike individual great auk bodies, the full contours of the event of its species extinction cannot be captured.

Since Ford regularly references his engagement with natural history illustration, it is worth mentioning that "Funk Island" and "The Witch" extend a lineage of great auk imagery that dates to Ole Worm's 1655 *Museum Wormianum*, which includes a detailed drawing of a single auk standing in profile (301). In this context, Ford's focus on anthropogenic violence directed at the great auk complicates a tradition that omitted from the frame human figures and the enabling conditions of natural history illustration (transit to specimen collection sites; the violence of specimen collection; the forms of patronage that supported specimen collection; and the creation of images, prints, and book). There is, however, another important framework for Ford's paintings: sixteenth-, seventeenth-, eighteenth-, and nineteenth-century written and visual narratives of marine bird and marine mammal hunting. Within this context, Ford's images dovetail those of his progenitors.[18] In other words, because human violence against nonhuman animals powers harvesting narratives, the inclusion of violence in Ford's work is not remarkable.

Both paintings trope the synecdoche of lastness, a common figure in extinction studies discourse. In lastness (or last-of-its-kind) narratives, one or a very diminished number of bodies stands in for larger populations and the socio-ecological attitudes that led to their species' demise. The paintings' different dimensions, historical moments, number of great auks portrayed, and types of violence comprise discrete approaches to lastness while also positioning it as durational. "The Witch" aligns with last of stories that traffic in elegy. Its lone great auk recalls other last animals, in particular Martha (passenger pigeon) and Lonesome George (Pinta Island tortoise), even while the painting's date lets the viewer know its great auk is not exactly the last. Though immobilized, the great auk's demeanor is calm; its head, beak, wing, and leg positions do not suggest struggle against its capture. Given the rope's presence, a viewer might read its gaze and outstretched wings as plaintive. The painting's size, shape, and tone make its content relatively intimate and thus accessible and digestible, especially relative to "Funk Island." Even though "The Witch" employs intertextuality, the scope of these gestures is more circumscribed than in "Funk Island" and derives from the following bio-geo-material circumstances. By 1840, great auk extinction was almost complete. The kinds of transatlantic travel and trade activities to which Funk Island was subjected had changed. Finally, St. Kildan harvesting practices and the source material for the painting are archipelagic not transatlantic; St. Kilda was not networked into the same travel and trade networks as Funk Island.

"Funk Island" tugs into view some different aspects of great auk lastness. It is much larger and more rectangular than "The Witch." The procession of great auks generates and augments this exaggerated horizontality. The painting's size, shape, and the mass of auks it presents operate in both microcosmic and proliferative fashion. The 42 great auks in "Funk Island" represent 0.00021% of its former colony's size.[19] While the great auks have a clear and final end point for their horrible horizontal procession, the fire, their point of origin is outside the frame's left edge. They could be emerging from the water or walking along other rocks. Paired with the mass of great auk bodies in the frame and their linear journey toward death, this absent origin invites imaginative consideration of the fantastic biomass of a full breeding colony. To assemble the 42 auks into the volume of their former colony would require replicating "Funk Island" 4,762 times. Lest this multiplication exercise seem ridiculous, part of what Ford's painting (and extinction studies more generally) requests is consideration of the weight – affective, biological, and existential – of great auk extinction.

The content and tone of "Funk Island" make avoiding this task difficult. Ford hammers the theme of the total truncation of great auk futurity directly via the unscathed northern gannet colony, the fire, the quote from Thomas, and the complete absence of great auk eggs.[20] The painting's subtitle, copulating human figures, and boats afloat in the background operate indirectly to further underline great auk finitude by indexing the transatlantic exploration and trade networks that ensnared Funk Island's auks. Without taking its source into account, the painting's subtitle works ironically. Clearly, burning great auks is not good conduct. And, just as clearly, because it was a repeated pattern that led to extinction, we know this conduct was not chastised. Taking its source into account elucidates the connection between the painting's subtitle, the excerpts from Thomas, and the copulating human figures. Ford sutures these details into a metonymic procession that underlines the consumptive nature of all metabolic activity. In other words, Ford underlines the fact that human pleasure, procreation, and population growth depend on the bodies of other creatures – in this case feathers for comfort during sex, and meat and eggs to fuel anthropogenic activities of harvesting North Atlantic cod stocks and expanding empires.

Ford's capacious nonhuman animal oeuvre presents persistent story templates for thinking the connections between colonialism, consumption, and extinction; individual paintings present "each time unique" stories that bring into view the precarious existence of and insist on the grievability of particular nonhuman animal bodies. Through the details in his paintings, Ford presents to and enlists the viewer in composing and processing an archive of the unfolding of the great auk's extinction. In part this archive is textual-material; it encompasses the visual texts and traditions of illustration Ford examines and critiques, the nonhuman animal bodies he studies in order to accurately draw his subjects, and long-term patterns of anthropogenic violence against the great auk. By bringing into view particular historical moments in great auk extinction, Ford's paintings, and the larger contexts they invoke and constellate, proliferate new moments for periodizing the Anthropocene and expanding the types of stories that get told under this rubric. By offering dates tied to the exploitation of nonhuman animal bodies, rather than dates primarily tied to human inventions, for instance the steam engine or atomic bomb, which regularly figure as potential golden spikes, Ford's work decenters more familiar forms of Anthropocene discourse. Just as crucially, Ford's great auk paintings make clear the importance of enlarging the archive of Anthropocene affects. For good reason, grief, melancholy, mourning, and sadness are the most often-invoked "negative" affects in extinction studies. Scholarship in this vein often aims to recuperate affective labor related to biodiversity loss. Ford's work takes a different tactic. He composes "Funk Island" and "The Witch" in ways that refuse elegy. Instead he presents horror, a lingering and non-recuperable response that adheres to the irrevocable severing of genetic lines, as the proper method for close(ly) reading extinction.

Notes

1 The estimate of the size of the Funk Island colony comes from Montevecchi. Other information in this paragraph is drawn from Birdlife; Fuller 34–49, 60–77, 80–85; Kirkham and Montevecchi; Kolbert 55–69; Montevecchi; and Montevecchi and Kirk.

2 My phrase "forms of violence" echoes van Dooren's "forms of life" but insists on the negative register in order to highlight that the lives of some depend on the deaths of others. Colebrook and Yusoff ("Aesthetic" and *Billion*) emphasize that life is fundamentally violent and requires the incorporative metabolism of others' bodies and labor.

3 Although anthropogenic factors were the primary causes of great auk extinction, an 1830 volcanic eruption that destroyed a notable Geirfuglasker (breeding station) off Iceland accelerated the species' end. After Funk Island was de-auked, this Geirfuglasker housed the only remaining sizable colony (Kolbert 62, Fuller 72). Its destruction drove the remaining great auks to the more accessible Eldey Island, which is where the last pair was captured and strangled (Fuller 80–85).

4 Thomas, quoted in Katz and Kazanjian 12, Fuller 65–66, and Kolbert 60.

5 For more about these forms of encounter and their timelines see Fuller 52–77, Gaskell, and Kolbert 55–69.
6 Whitbourne, quoted in Fuller 60, Kolbert 60, and Montevecchi and Kirk. Other exploration narratives that mention the great auk include Cartier, Denys, Fabricius, Hakluyt, Martin, and the aforementioned Thomas. Standing reserve is from Martin Heidegger's "The Question Concerning Technology."
7 See Ford "Condemned" and "Dying" (Carolina parakeet), "American" (flamingo), "Funk" and "Witch" (great auk), "Falling" and "Visitation" (passenger pigeon), and "Benjamin's" and "Our" (turkey). For Audubon's illustrations see "American" "Carolina," "Great," "Passenger" "Wild I," and "Wild VI."
8 Lewis and Maslin offer a history of the Anthropocene concept. Tomkins's in-depth profile of Ford is wide-ranging. For representative interviews, see Bendrick, "Interview," Kazanjian, and "Political." Arning, Buford, Burroughs, Finkelstein, Jacobson, Jakub, Katz, Larson, Ollman, and Platt have penned short essays or exhibition reviews. See Traisnel and Whittle for scholarly articles on Ford.
9 See especially DeLoughrey 1–32 for her theorization of the idea of provincializing the Anthropocene.
10 I make this point this way to underline all creatures' fundamental dependence on the earth and its systems.
11 See Clark for an extended discussion about Earth's instability. My point about human–earth relations vis-à-vis reading and writing aligns with and extends Szerszynski 165. On reading and/in the Anthropocene see also Menely and Taylor.
12 See Bladow and Ladino for essays about environmental affect; my chapter discusses form, affect, and the Anthropocene (Merola "'what'"). I also address forms of Anthropocene affect in "Materializing" and "Mediating." Albrecht considers "earth emotions" from the perspective of environmental philosophy.
13 Kalshoven and Lorimer offer other environmental humanities approaches to the great auk.
14 One aim here is to emphasize productive tension between viewing, reading, and writing, which are related but distinct sensorial, cognitive, and interpretive practices. Another is to highlight how description can function as witnessing, as a form of reading that refuses to occlude what the text contains.
15 Garefowl is from the Norwegian *geirfugl*. Carolus Linnaeus assigns the binomial *Alca impennis* in *Systema Naturae* (1758, vol. 1, 130).
16 The full Thomas passage is included on Katz and Kazanjian 12, the page facing the reproduction of "Funk Island." See also footnote 4, above.
17 Quoted in Fuller 75. See also Gaskell 142.
18 Fuller's text exemplifies the process of assembling an archive of great auk extinction and includes many notable great auk images; see especially 22, 26, 35, 46, 47, 53, 62, 67, 70, 92, 98, 99, and 102. Partridge uses Audubon as a case study to delineate conventions in ornithological illustration. Fuller includes images of great auk harvesting on 64 and 65.
19 This point takes Montevecchi's estimate of 200,000, mentioned in the opening, as accurate.
20 Since breeding, incubating eggs, and raising young are the only reasons for great auks to be on land, if biological accuracy was the painting's point it would include eggs.

Works cited

Albrecht, Glenn A. *Earth Emotions: New Words for a New World*. Cornell UP, 2019.

Arning, Bill. "Walton Ford's Non-Fiction Fiction." *Brutal Beauty: Paintings by Walton Ford*, Bowdoin College Museum of Art, 2000, pp. 16–28.

Audubon, John James. "American Flamingo. Plate CCCCXXXI." *The Birds of America*, Vols. I–V. Double elephant folio. Aquatints by Robert Havell. Full original hand-color by R. Havell. London, J. J. Audubon, 1827–1839.

———. *The Birds of America*, Vols. I–V. Double elephant folio. Aquatints by Robert Havell. Full original hand-color by R. Havell. London, J. J. Audubon, 1827–1839.
———. "Carolina Parrot. Plate XXVI." *The Birds of America*, Vols. I–V. Double elephant folio. Aquatints by Robert Havell. Full original hand-color by R. Havell. London, J. J. Audubon, 1827–1839.
———. "Great Auk. Plate CCCXLI." *The Birds of America*, Vols. I–V. Double elephant folio. Aquatints by Robert Havell. Full original hand-color by R. Havell. London, J. J. Audubon, 1827–1839.
———. "Passenger Pigeon. Plate LXII." *The Birds of America*, Vols. I–V. Double elephant folio. Aquatints by Robert Havell. Full original hand-color by R. Havell. London, J. J. Audubon, 1827–1839.
———. "Wild Turkey. Plate I." *The Birds of America*, Vols. I–V. Double elephant folio. Aquatints by Robert Havell. Full original hand-color by R. Havell. London, J. J. Audubon, 1827–1839.
———. "Wild Turkey. Plate VI." *The Birds of America*, Vols. I–V. Double elephant folio. Aquatints by Robert Havell. Full original hand-color by R. Havell. London, J. J. Audubon, 1827–1839.
Bendrick, Lou. "Walton Ford Brings Testosterone to Nature Painting." *Grist*, 19 September 2006, https://grist.org/article/bendrick2/. Accessed 20 November 2020.
BirdLife International. "*Pinguinus impennis. The IUCN Red List of Threatened Species 2016*: e.T22694856A93472944." *IUCNRedList.org*, https://dx.doi.org/10.2305/IUCN.UK.2016-3.RLTS.T22694856A93472944.en. Accessed 1 October 2016.
Bladow, Kyle and Jennifer Ladino, editors. *Affective Ecocriticism: Emotion: Embodiment, Environment*. U Nebraska P, 2018.
Brannen, Peter. *The Ends of the World: Volcanic Apocalypses, Lethal Oceans, and Our Quest to Understand Earth's Past Mass Extinctions*. Ecco, 2017.
Buford, Bill. "Field Studies: Walton Ford's Bestiary." *Pancha Tantra*, Taschen, 2009, pp. 8–11.
Burroughs, Franklin. "Visions and ReVisions: John James Audubon and Walton Ford." *Brutal Beauty: Paintings by Walton Ford*, Bowdoin College Museum of Art, 2000, pp. 5–15.
Cartier, Jacques. *The Voyages of Jacques Cartier: Published From the Originals With Translations, Notes and Appendices*. (1534). Translated and edited by Henry Percival Biggar. Ottawa: F.A. Acland, 1924.
Clark, Nigel. *Inhuman Nature: Sociable Life on a Dynamic Planet*. London: Sage, 2011.
Colebrook, Claire. *Death of the Posthuman: Essays on Extinction, Vol. 1*. Open Humanities Press, 2014.
———. "Framing the End of the Species: Images Without Bodies." *Symploke*, vol. 21, no. 1–2, 2013, pp. 51–63.
———. *Sex after Life: Essays on Extinction, Vol. 2*. Open Humanities Press, 2014.
DeLoughrey, Elizabeth M. *Allegories of the Anthropocene*. Duke UP, 2019.
Denys, Nicolas. *The Description and Natural History of the Coast of North America (Acadia)*. (1672). Translated by W. F. Ganong. Toronto: The Champlain Society, 1908.
Ellis, Richard. *No Turning Back: The Life and Death of Animal Species*. Harper, 2004.
Fabricius, Otto. *Fauna Groenlandica: systematice sistens animalis Groenlandiae occidentalis hactenus indagata, quoad nomen specificum, triuiale, vernaculmque: synonyma auctorum plurium, descriptionem, locum, victum . . .* Hafniae et Lipsiae: Ioannis Gottlob Rothe, 1780.
Ford, Walton. "American Flamingo." 1992, *Pancha Tantra*. Taschen, 2009, p. 25.
———. "Benjamin's Emblem." 2000, *Pancha Tantra*. Taschen, 2009, p. 131.
———. "Condemned." 2007, *Pancha Tantra*. Taschen, 2009, p. 269.

———. "Dying Words." 2005, *Pancha Tantra*. Taschen, 2009, pp. 238, 240–41.
———. "Falling Bough." 2002, *Pancha Tantra*. Taschen, 2009, pp. 169, 170–72.
———. "Funk Island." 1998, *Pancha Tantra*. Taschen, 2009, pp. 102–03.
———. "Interview with the Artist, April 1997." *Walton Ford*. Southeastern Center for Contemporary Art, 1997, pp. 15–18.
———. "Our Emblem." 1992–1993, Private Collection. "SALE 2793: Post-War and Contemporary Afternoon Session." *Christie's*, www.christies.com/lotfinder/drawings-watercolors/walton-ford-our-emblem- 5739615-details.aspx?intobjectid=5739615&lid=1&sc_lang=en. Accessed 10 January 2018.
———. *Pancha Tantra*. Taschen, 2009.
———. "Visitation." 2004, *Pancha Tantra*. Taschen, 2009, p. 227.
———. "The Witch of St. Kilda." 2005, *Pancha Tantra*. Taschen, 2009, pp. 264–65.
Fuller, Errol. *The Great Auk*. Harry N. Abrams, 1999.
Gaskell, Jeremy. *Who Killed the Great Auk?* Oxford UP, 2000.
Grusin, Richard, editor. *After Extinction*. U of Minnesota P, 2018.
Hakluyt, Richard. *The principal navigations, voyages, traffiques & discoveries of the English nation made by sea or over-land to the remote and farthest distant quarters of the earth at any time within the compasse of these 1600 Yeeres*. Glasgow: James MacLehouse and Sons, 1904.
Harvie-Brown, John Alexander and Thomas Edward Buckley. *A Vertebrate Fauna of the Outer Hebrides*. D. Douglas, 1888.
Heidegger, Martin. "The Question Concerning Technology." *Martin Heidegger: Basic Writings*, edited by David Farell Krell, Harper & Row, 1977, pp. 287–317.
Heise, Ursula K. *Imagining Extinction: The Cultural Meanings of Endangered Species*. The U of Chicago P, 2016.
Jacobson, Mark. "Nature Boy." *New York*, 21 October 2002. https://nymag.com/nymetro/arts/art/n_7839/. Accessed 10 January 2018.
Jakub, Lucy. "Walton Ford: Twenty-First Century Naturalist." *The New York Review of Books*, 16 December 2018. www.nybooks.com/daily/2018/12/16/walton-ford-twenty-first-century-naturalist/. Accessed 12 September 2019.
Kalshoven, Petra Tjitske. "Piecing Together the Extinct Great Auk: Techniques and Charms of Contiguity." *Environmental Humanities*, vol. 10, no. 1, 2018, pp. 150–70.
Katz, Steven. "Walton Ford: A Personal History of His Work." *Walton Ford: Tigers of Wrath, Horses of Instruction*, edited by Steven Katz and Dodie Kazanjian, Harry N. Abrams, Inc., 2002, pp. 6–11.
Katz, Steven and Dodie Kazanjian. *Walton Ford: Tigers of Wrath, Horses of Instruction*. Harry N. Abrams, Inc., 2002.
Kirkham, I. R. and W. A. Montevecchi. "The Breeding Birds of Funk Island, Newfoundland: An Historical Perspective." *American Birds*, vol. 36, no. 2, 1982, pp. 111–18.
Kolbert, Elizabeth. *The Sixth Extinction: An Unnatural History*. Henry Holt and Company, 2014.
Larson, Barbara. "Ornithology and Allegory: Walton Ford." *Antennae: The Journal of Nature in Visual Culture*, vol. 25, 2013, pp. 27–32.
Lesjak, Carolyn. "Reading Dialectically." *Criticism*, vol. 55, no. 2, 2013, pp. 233–77.
Lewis, Simon L. and Mark A. Maslin. "Defining the Anthropocene." *Nature*, vol. 519, 2015, pp. 171–80.
Linneaus, Carolus. *Systema naturae per regna tria naturae, secundum classes, ordines, genera, species, cum characteribus, differentiis, synonymis, locis. Tomus I. Editio decima, reformata*. Holmiae: Laurentii salviii, 1758.
Lorimer, Jamie. "On Auks and Awkwardness." *Environmental Humanities*, vol. 4, 2014, pp. 195–205.

Martin, Martin. *A late voyage to St. Kilda, the remotest of all the Hebrides, or the Western isles of Scotland with a history of the island, natural moral, and topographical: wherein is an account of their customes religion, fish, fowl, &c.: as also a relation of a late imposter there, pretended to be sent by St John Baptist, by M. Martin, gent.* London: D. Brown and T. Goodwin, 1698.
Menely, Tobias and Jesse Oak Taylor. "Introduction." *Anthropocene Reading: Literary History in Geologic Times*, edited by Tobias Menely and Jesse Oak Taylor, The Pennsylvania State UP, 2017, pp. 1–24.
Merola, Nicole M. "Materializing a Geotraumatic and Melancholy Anthropocene: Jeanette Winterson's *The Stone Gods*." *The Minnesota Review*, vol. 83, 2014, pp. 122–32.
———. "Mediating Planetary Attachments and Planetary Melancholy: Lars von Trier's *Melancholia*." *Design, Mediation, and the Posthuman*, edited by Amy Propen, Colbey Emmerson Reid, and Dennis Weiss, Lexington Books, 2014, pp. 249–67.
———. "'What do we do but keep breathing as best we can this/minute atmosphere": Juliana Spahr and Anthropocene Anxiety." *Affective Ecocriticism: Emotion, Embodiment, Environment*, edited by Kyle Bladow and Jennifer Ladino, U of Nebraska P, 2018, pp. 25–49.
Montevecchi, William. "The Great Auk Cemetery." *Natural History*, vol. 103, no. 8, 1994, p. 6.
Montevecchi, William A. and David A. Kirk. "Great Auk (Pinguinus impennis), version 2.0." *The Birds of North America*, edited by A. F. Poole and F. B. Gill, Cornell Lab of Ornithology, 1 January 1996. https://doi.org/10.2173/bna.260. Accessed 10 September 2019.
Partridge, Linda Dugan. "By the Book: Audubon and the Tradition of Ornithological Illustration." *Huntington Library Quarterly*, vol. 59, nos. 2 & 3, 1998, pp. 269–301.
Platt, Ron. "The Legacy of Empire." *Walton Ford*. Southeastern Center for Contemporary Art, 1997, pp. 4–6.
"Political Humor and Colonial Critique: Walton Ford." *art21*. https://art21.org/read/walton-ford- political-humor-and-colonial-critique/. Accessed 20 September 2019.
Quammen, David. *The Song of the Dodo: Island Biogeography in an Age of Extinction*. Scribner, 1996.
Rose, Deborah Bird. *Wild Dog Dreaming: Love and Extinction*. U Virginia P, 2011.
Rose, Deborah Bird and Thom van Dooren, editors. "Unloved Others: Death of the Disregarded in the Time of Extinctions." *Australian Humanities Review*, vol. 50, 2011.
Rose, Deborah Bird, Thom van Dooren, and Matthew Chrulew, editors. *Extinction Studies: Stories of Time, Death, and Generations*. Columbia UP, 2017.
Schuster, Joshua. "Photographing the Last Animal." *Antennae: The Journal of Nature in Visual Culture*, vol. 41, 2017, pp. 102–22.
Sodikoff, Genese Marie. *The Anthropology of Extinction: Essays on Culture and Species Death*. Indiana UP, 2012.
Szerszynski, Bronislaw. "The End of the End of Nature: The Anthropocene and the Fate of the Human." *Oxford Literary Review*, vol. 34, no. 2, 2012, pp. 165–84.
Thomas, Aaron. *The Newfoundland journal of Aaron Thomas, Able Seaman in H.M.S. Boston: a journal written during a voyage from England to Newfoundland and from Newfoundland to England in the years 1794 and 1795, addressed to a friend*. Edited by Jean M. Murray. London: Longmans, 1968.
Tomkins, Calvin. "Man and Beast." *The New Yorker*, 26 January 2009, pp. 50–57.
Traisnel, Antoine. "Huntology: Ontological Pursuits and Still Lives." *Diacritics*, vol. 40, no. 2, 2012, pp. 4–25.
van Dooren, Thom. *Flight Ways: Life and Loss at the Edge of Extinction*. Columbia UP, 2016.
Walton Ford. Southeastern Center for Contemporary Art, 1997.

Weisman, Alan. *The World Without Us*. Thomas Dunne Books, 2007.
Whitbourne, Captain Richard. *A discourse and discovery of New-found-land, with many reasons to proove how worthy and beneficiall a plantation may there be made, after a better manner than it was. Together with the laying open of certaine enormities and abuses committed by some that trade to that countrey, and the means laid downe for reformation thereof: As also a loving invitation: and likewise the copies of certaine letters sent from that countrey; which are printed in the latter part of this booke*. London: Feliz Kingston, 1623.
Whittle, Matthew. "Lost Trophies: Hunting Animals and the Imperial Souvenir in Walton Ford's *Pancha Tantra*." *The Journal of Commonwealth Literature*, vol. 51, no. 2, 2016, pp. 196–210.
Wolfe, Cary. "Condors at the End of the World." *After Extinction*, edited by Richard Grusin, U of Minnesota P, 2018, pp. 107–22.
———. "Each Time Unique: The Poetics of Extinction." *Animalities: Literary and Cultural Studies Beyond the Human*, edited by Michael Lundblad, Edinburgh UP, 2017, pp. 22–42.
Worm, Ole. *Museum Wormianum, seu, Historia rerum rariorum: tam naturalism, quam artificialium, tam domesticarum, quam exoticarum, quae Hafniae Danorum in aedibus authoris servantur*. Lugduni Batavorum: Apud Iohannem Elsevirium, 1655.
Yusoff, Kathryn. "Aesthetics of Loss: Biodiversity, Banal Violence and Biotic Subjects." *Transactions of the Institute of British Geographers*, vol. 37, 2012, pp. 578–92.
———. *A Billion Black Anthropocenes or None*. U of Minnesota P, 2018.

4

BARTLEBY AND THE POLITICS OF MEASUREMENT

Helena Feder

> Take five-and-twenty heaps of cinders dumped here and there in an outside city lot; imagine some of them magnified into mountains, and the vacant lot the sea; and you will have a fit idea of the general aspect of the Encantadas, or Enchanted Isles. A group rather of extinct volcanoes than of isles; looking much as the world at large might, after a penal conflagration.
>
> – *Sketch First, "The Encantadas, or Enchanted Isles"*

"Man" is not a reliable narrator. He is a narrator who has taken himself as the measure of all things.[1]

The Anthropocene seems, quite literally, another such measure. What might it mean to read closely, to discern meaning or value – one might say weigh or measure – in an age at once characterized by a new awareness of vertiginous scale and complex interconnection and by the collapse of multispecies agency into a singularity, an idea of "Man?" The world is, of course, more-than-human and more than our measure of it. Scale itself expands and recedes almost infinitely while "close" has become uncomfortably closer – from toxins inherited in utero[2] to cultivated misinformation sweeping the globe faster than any disease[3] – and the human future now often seems behind us. In the above passage from Melville's "The Encantadas, or Enchanted Isles," cinder heaps in an exurban lot become burnt-out volcanos protruding from a strangely passive sea. This land-and-seascape, in turn, becomes an image of the world consumed by a "penal conflagration" (126). Reading it in this moment, it calls to mind every ecosystem struggling with increasingly catastrophic blazes and the sixth mass extinction itself, threatening to turn much of the living world into a vacant gray lot.

Close reading (in) the Anthropocene need not be yet another measure of *homo sapiens* as the *only* animal that weighs and knows. It may also give rise to the reverse, and to our growing awareness of transspecies vulnerability. If "Man," the Man of empire and industry, is an unreliable narrator save of his own fears and desires, then "Bartleby, the Scrivener: A Story of Wall Street,"[4] perhaps the most scrutinized story in American literature, is an apt choice for a close reading in and of the Anthropocene. However one dates the onset of this geologic era,[5] the designation of the Anthropocene suggests at once the scope and severity of anthropogenic damage to the world, the material significance of *homo sapiens*, and the way in which we are still *in*significant, evolutionarily interwoven. That is, in the emerging understanding of deep geological time, we, like Bartleby, are not so "particular," not so alone. As Nigel Clark contends, scientific research increasingly reveals a long history of climactic and planetary volatility, evidence of periodic revolutions; while the idea of the Anthropocene is often read by "critical social scientists to be supportive of their case for an ultimate 'end of nature,' in the earth sciences it is more likely to be read as an affirmation that our species belongs among other biological and geophysical forces" (*Inhuman Nature* 15). Quoting Alain Badiou, he warns against thinking we're done with the ground; while we must do everything possible to change human behavior, "most of material reality is not ours to make over . . . we must also account for forces, events and objects that can't be done differently or done away with – or things that will be otherwise whatever *we* chose to do. These too are political questions" (*Inhuman Nature* 18). The inhuman Earth is not only constitutive of our humanity but also the ground of ethics and politics, the foundation of this close and distant reading of Melville's much-read story.

By distant I mean preoccupied by the shared and contested meaning of the present moment of global ecological change. In *Ecocriticism on the Edge*, Timothy Clark argues that the Anthropocene "names a newly recognized context that entails a chastening recognition of the limits of cultural representation as a force of change in human affairs, as compared to the numerous economic, meteorological, geographical and microbiological factors and population dynamics, as well as scale effects" (21). In these times, "the scalar gap" between what can be seen and what is really happening "increasingly feeds into a panic syndrome that can be nicknamed 'Anthropocene disorder' . . . a state in which "the mind is suspended, uncertainly, between a sense of rage and even despair on one side, and a consciousness of the majority perception of such reactions as disproportionate and imbalanced on the other" (140). In a frame of contradiction that's become all too familiar, behind this "break-down in the sense of proportion and of propriety when making judgements" is the human species as "the particular site of a capitalization of information and energy far speedier than in the rest of nature, making it a capricious accelerator and changer of Earth system processes" (145, 150).[6]

While "discrepancy of scale is crucial to defining the eventhood of the Anthropocene as a threshold concept," the "predatory supremacy of global neoliberal capitalism would represent a further, exploitative intensification of this scalar disjunctiveness . . . as a way of gaining power over human and nonhuman others" (*Ecocriticism* 151). But there is still more to account for here, including what is problematic in the idea of the Anthropocene itself as a means to critique the exploitative intensification it identifies – from, as Gary Snyder once said, blaming the entire species for the errors of white male capitalists[7] to, as Stacy Alaimo points out, the problems of transcendence posed by certain forms "zooming out" to conceptualize or depict anthropogenic impact.[8] In space and time, this neoliberal, technological intensification is the outcome of a slow, sprawling violence; as Kathryn Yusoff has it in *A Billion Black Anthropocenes or None*, the Anthropocene might seem "a dystopic future that laments the end of the world, but imperialism and ongoing (settler) colonialism have been ending worlds for as long as they have been in existence. The Anthropocene . . . is just now noticing the extinction it has chosen to continually overlook in the making of its modernity and freedom" (xiii). And yet, she argues, the term may be read dialectically, as containing "a form of Anthropogenesis – a new origin story and ontics for man – that radically rewrites material modes of differentiation and concepts of life, from predominantly biopolitical notions of life toward an understanding of life's geophysical origination," suggesting that "a more nuanced notion of 'geological life'" may challenge "the construction of the Anthropocene as an undifferentiated social stratification" ("Anthropogenesis . . ." 3).

In other words, even as the Anthropocene may serve to reinscribe human as Man and Man as separate from nature, as a measure or end in himself, obfuscating "climate racism, social injustice in fossil fuels, and differentiated histories of responsibility through homogenization in a 'we' of the Anthropocene," it may also serve as a shift "from humanist thought which characterized the inhuman as a dehumanizing force, to a concept of the inhuman as materially constitutive of the possibilities of life" ("Anthropogenesis . . ." 7). Echoing Marx's lyrical first sentence of his *Manifesto*, Yusoff characterizes the Anthropocene as that which "both names the *spectre* of a fossil-fueled geologic life that haunts the present (while leaving it unexamined in geopolitical terms) *and* opens up an epochal rift for speculative contemplation that extends well beyond industrialization and capitalism into evolutionary futures" ("Anthropogenesis . . ." 5, first italics mine). We might read this as a formulation of the geist in which "Man" seems to become his own ghost, an uninvited guest that may provoke a consideration of the ethics of inhabitation in human and inhuman, coevolutionary and co-constitutive, terms.

In the annals of American literature "Bartleby, the Scrivener" is *the* story of an uninvited guest. It dramatizes the bad faith of humanism as the bad faith of capitalism and, in the context of close reading the Anthropocene, invites speculation about agency, islands, and becoming rock. Indeed, Bartleby is the very thing that "will be otherwise whatever *we* chose to do" as Nigel Clark says of the geological world, the very thing that may prefer not to be changed or moved. He is a human being that personifies inhuman agency, I will argue, though it might seem tempting to pathologize Bartleby as so many have done before (as schizophrenic, depressed,[9] autistic,[10] agoraphobic,[11] alienated, anorexic, or entropic[12]), in this case as a figure anticipating the Anthropocene disorder, a Pinocchio inversion, a boy-turned-wood. Or, more to the point, a man petrified. But even reading this contemporary phenomenon back into the industrial and ecological past of the 1850s, it would be the Lawyer–narrator, not Bartleby, who suffers from a "break-down in the sense of proportion and of propriety when making judgements."

Along these lines one could see Bartleby as figure for humanity writ large, *en masse* an inhuman, geologic force. And it is also possible to read Bartleby as a figure of the 99%, against the false, depoliticizing universality of humanism, as Russ Castronovo does in his analysis of the Occupy movement's analogous rendering of Bartleby, which leads him to articulate analogy itself as a system that "place[s] things in relation to one another, pointing to common structures that allow for extrapolation" (266).[13] Perhaps it would, as both Dan McCall and Elizabeth Hardwick suggest, dishonor Bartleby to think of him as something or "someone else" (as Henry David Thoreau, Nathaniel Hawthorne, or Melville himself) but, given the incredible amount of scholarship and theory generated from the story, does it make sense to argue that "Bartleby" is "not an elaborate code for the modern critical mind" (76, 97)? *Yes*, but only if by "code" we mean, as McCall does, something which admits of a key, something which must be cracked to reveal "the 'real' story underneath" (97). Here I will suggest a reading with, rather than of Bartleby, one as naive as McCall might wish for, one which sees him cast as the ground of meaning-making. What could be more naive than seeing such ground as *the ground beneath our feet*? While I do not argue the Lawyer is a stand-in for the reader (as McCall does), he is a reliable narrator of "Man's" unreliability, a measure of the humanist fantasy of detachment sometimes called individuality, sometimes independence, sometimes freedom. "Bartleby" is at its core a story of two persons and parts; here, this duality is the ground of ethics and of politics: the interdependence of self *and* other (be it human, nonhuman, or inhuman).[14]

The literal ground of "Bartleby" is Manhattan, an island symbolic of American individualism and exploitation,[15] a mythos that partakes of humanism's old discourse of worlds as islands and humans as islands. It is a narrative animated by a desire for mastery if not management, a levee for the fear of the nonhuman and inhuman; in this narrative, the island is a model for containment and control. But the island is also a model of interconnectedness and a key source of ecological knowledge, a place surrounded in the flux of the known and unknown, the complex intra-active air-and-water systems, the real and symbolic "currency" of anthropogenic damage.[16] Like the idea of the Anthropocene, the island turns two ways and, as ecogeological metaphor, may provide a way into thinking about the ethics of living in the midst of difference and damage, and the politics of measurement in the polis of the planetary commons.

The biographical evidence for Melville's material, global consciousness, from his studies to his travels, is well known. "Melville's odyssey through open oceans, expansive prairies, frozen extremes, and verdant islands exposed him to a 'realizing sense,'" Timothy Marr argues, "of the immense power and terror of the material earth in space and time" (187), that is, of the earth, from the immensity of the heavens to the depths of the oceans, as a "'loose fish' in deep time" (188). Melville's geological reading and firsthand experience of avalanches, earthquakes, and volcanic eruptions "undermined the earth's solidity. The geological cataclysms of the planet's chaotic evolution rendered destruction and waste as the formative processes of the earth's own foundation" (Marr 194). It undermined, I would argue, a sense of the earth as a place made for humans. But in Melville's writings, it is the ocean, "the fluid and unfathomable abyss of the ocean that most embodied matter's cannibalistic power to devour human lives" (Marr 195).

But this cannibalistic power is also a first principle of life, a key source of agency in the world.[17] Melville's oceanic imagination encompasses the torque of the sublime, the two turns in which we twist. Just as, in an overlapping frame, Melville registers the duality of interconnectedness. As Charles Waugh argues, while transcendentalist writers emphasized "the important connection between the local and global, the particular and the universal," Melville "actually demonstrates how transcendentalism, a philosophy about the hidden connections between humans and things, in practice becomes globalization, a process in which people, production, travel, nature, and commerce are ever increasingly interconnected" (208). A global consciousness that "manifests itself in the city not only as a hellish place . . . but also as the locus for a kind of global economics capable at times of dominating other aspects of life" (Waugh 223). As the epigraph of this chapter suggests, imagery in the first sketch of "The Encantadas" echoes the urban inhumanity of "Bartleby." Again, while to Melville the Galapagos resemble "five-and-twenty heaps of cinders dumped here and there in an outside city lot . . . magnified into mountains, and the vacant lot the sea" (126), Manhattan can seem as barren of human feeling as these isles (perhaps more so, given what Melville found there).[18] Indeed, it is reasonable to suppose that Melville's time at sea in the early 1840s, particularly his time in the Galapagos, was in his thoughts while he wrote "Bartleby," given the first installment of "The Encantadas" followed in *Putnam's* just three months later. While Melville may or may not consistently depict humans as "islands of decreasing entropy amid the increasing entropy of the universe,"[19] Bartleby is certainly such an island. An island on an island, he prefigures the tortoise of "The Encantadas," emblematic of "uninhabitableness" (126) and dogged, even suicidal, determination: "I have known them in their journeyings ram themselves heroically against rocks. . . . Their crowning curse is their drudging impulse to straightforwardness in a belittered world" (132). The same might be said of humanity. Indeed, the narrator even imagines the "ghost of a gigantic tortoise, with 'Memento ****' burning in live letters upon his back" (129).[20]

Our Lawyer's strange, even deranged,[21] Manhattan story, his island narrative of a visitant copyist, a stranger isolated and isolating, *seems* the story of his haunting. A haunting of Wall Street by Wall Street. In the long course of what has come to be called "the Bartleby Industry" (first by McCall), Bartleby's oft-quoted "preference not to" has evolved from a phrase of philosophical significance for literary criticism to the subject of philosophy itself read, alternately, as an example of radical contingency (Agamben), a passivity of radical neutrality (Blanchot), an "agrammatical formula" for undermining the social contract (Deleuze), an expression of *the* ethical dilemma or the dilemma of ethics (Derrida), an absolute refusal (Hardt and Negri), or an affirmation of irreducible negativity (Žižek).[22] In their different ways, these readings, however material, seem to keep Bartleby ghostly, a presence of detached signification, collapsing or sidelining his radical embodiment, his intransigent inhabitation, in this catchy catchphrase. From the first, it was the Lawyer's desire to "cheaply purchase delicious self-approval" (23), his framing of self-worth as a consumable product, that stood behind and opposite his characterization of Bartleby as a "wight" (25), "a very ghost" (25), and "cadaverous" (27), and yet it is he who felt insubstantial in the face of Bartleby's "singular" being, his "passive resistance" (23). In the following rendering of the story, which close reads the Lawyer's reading of Bartleby as an island, I mean to suggest he is more than his phrase and more than metaphor. To sketch a few lines between near and far in these times of scalar confusion, this close reading of "Bartleby" is a distant skipping of stones: Bartleby as rock, as island, and island as material reality.

Ah, Bartleby

Bartleby appears, uncannily motionless, on the threshold of a law office hemmed-in on all sides by taller buildings, with some natural light but no views, described by the narrator as "deficient in what landscape painters call life" and "a huge square cistern" (14). The detailed illustration of the physical setting and population of his office and, in particular, the Lawyer's use of the word "cistern" (turning a round peg into a square hole) at the beginning of the tale situates it, and him, in a very particular way.[23] If this office is a reservoir, an internal ocean of sorts, then Bartleby is its lone rock, an island – on Wall Street, an island within an island. Or, an island thrice: an island embodied, a cistern himself. In the figure of Bartleby, water and rock, rock and water, encompass and compose each other.[24] For our narrator is, himself, still water, the "eminently *safe* man" who believes "the easiest way of life is the best" (14), for whom the "advent of Bartleby" (15) is an epoch of catastrophe.

At first, Bartleby is like a stone rolling down a steep hill: a force of growing momentum, he is inordinately productive day and night, "gorge[ing]" (19) himself on documents. But soon, with "a gray eye dimly calm" (20), he refuses to check copy, eventually refusing to copy at all. In refusing the endless repetition of the scrivener, Bartleby explicitly trades the place of Sisyphus for that of his rock. Repeating only his resistance to repetition as a positive preference, Bartleby's stolid inhumanity is like the "dead brick wall" of the office at which he continually stares that, in the end, becomes the wall of the Tombs against which he dies. Radically different but, to the Lawyer's eyes, damaged,[25] Bartleby's *sediment* arouses his curiosity, consternation and, finally, *sentiment*: the Lawyer's assertion of "a bond of common humanity" (28). But this sentiment cannot take hold, cannot take root in Bartleby; faced with his radical difference, his inanimate solidity, it turns, becomes its "opposite:" the Lawyer's "melancholy merge[s] into fear . . . pity into repulsion" (29). Finding Bartleby immobile as rock, "prefer[ring] to give no answer" and "prefer[ring] not to be a little reasonable" (30), the Lawyer tries to cast off this "millstone" (32) by removing his office from the premises, leading to Bartleby's removal to the Tombs, his death, and the narrator's exclamation as an uncanny *apostrophe*: "Ah Bartleby! Ah humanity!" (45).

By becoming more pallid, impassive, and immobile than the narrator's "plaster-of-paris bust of Cicero" (21), by refusing to become the pliable subject of our narrator's biography, Bartleby becomes solidly inhuman, a rock rooted in place, a mountain, an island. Indeed, plaster-of-paris is calcinated gypsum, a sedimentary rock! And a rock formed from the oceans.[26] "[Mo]tionless" and "singular[ly] sedate" (19), Bartleby reverses the fundamental background of humanism, the idea that the ground itself, the earth, is the static backdrop for human activity. Bartleby's "irreversible" decision (22) *not to* functions as a decision to reverse the assumption of Earth's passive acceptance of our actions. A refusal that, at one point, turns the narrator "into a pillar of salt" (21) and, later, causes him to "stagger in his own plainest faith" (22) – like Voltaire after the Lisbon earthquake.[27] In a story that figures humans as co-constituted by nature and culture, or matter (such as the connection between the "ugly moods" of Turkey and Nippers and their habits of consumption)[28] *and* the immaterial (such as the infectious nature of the word "prefer"), Bartleby, by becoming "a fixture" (32) with the "pallid haughtiness" (28) of an Alp, by making background foreground, by refusing his humanist history, reminds us that *"Man" is not a reliable narrator.*

If Wall Street is an embankment against the vicissitude of fate, against the ocean of material mystery and uncertain futurity, Bartleby is the embodiment of solitude that dramatizes the deep material interconnectedness of what's been categorized as animate and inanimate, an island that situates human need and its satisfaction in a multifarious world, not as origin and end, but as part of a continuous and contingent commons. As metaphor, the island reminds us that no "Man," no individual, no organism is an island; indeed, no island is an island; we all arise from the magma and crust of the earth. Agency, life, is shifting, messy, and mutually constituted. Intimately inhabiting the ecogeologic, our seemingly "contained" bodies contain the sea that envelops and connects us all.

While Bartleby's refusal to perform, his preference that so mystifies his master, is, of course, a refusal of the subject to be constituted by industrial capitalism, the story as a whole dramatizes the stratified nature of the continuum of freedom and contingency, as when the Lawyer asks Bartleby, "What *earthly* right have you to stay here? Do you pay any rent? Do you pay my taxes? Or is this property yours?" (35, italics mine). But is Bartleby's "wondrous ascendency" and "cadaverous triumph" (35) over the narrator an uncanny "return" of the geological to the consciousness of the biological? The Earth, treated inhospitably in the construction of the human, become inhospitable? The story is, at least, an insistence on the necessity of thinking further about difference and the politics of measurement, including the measure of the Lawyer's "common humanity."[29]

In, or against, the service of "common humanity," a number of critics have read Bartleby as the Lawyer-narrator's double (such as Mordecai Marcus in "Melville's Bartleby as a Psychological Double" and Marvin Fisher in "'Bartleby' Melville's Circumscribed Scrivener"), and some have read Bartleby as the antithesis of the narrator (such as Michael Paul Rogan in "Melville and the Slavery of the North"). In *American Terror*, Paul Hurh argues that the lawyer's anxiety "determines Bartleby's originality through differentiation, by measuring out the distance between our internally divided selves and this specter of potentiality, of raw and unconditioned freedom" (222).

It is equally true (and here, important) that the narrator *is Bartleby's double*; that is, the narrator, infected by Bartleby's presence and "preferences," becomes like him, the "echo" of Bartleby's formula and its echo chamber. Again, where Bartleby is compared to the plaster-of-paris bust of Cicero, our narrator is – as he tells us just a few paragraphs later – "turned into a pillar of salt" (21). The narrator's own becoming rock makes that final apostrophe seem not simply or only an equation of Bartleby with humanity but a suggestion of humanity's final fate as rock. We are, of course, not only reading the narrator reading Bartleby, but close reading the narrator reading himself through Bartleby. Both Bartleby and our Lawyer–narrator are mutable (according to Anders Gullestad, this is what Deleuze misses, the mutability of the formula and fact of the narrator).[30] That is, mutable until ossification.

The pillar of the community becomes a pillar of salt. Does he turn to look back at, or turn into, something which he shouldn't? Or does he merely recognize the inhumanity on which his philanthropic impulses rest? It seems only right, now, to nod to the similarities, the echoes between – though not the equation of – our Lawyer and Ishmael, and Bartleby and Ahab. And, as doubles of each other, our narrator and Ahab, Bartleby and the white whale (or, as Leo Marx suggests, Bartleby and Ishmael).[31] These doubles are, oddly, both balanced and imbalanced. As Beongcheon Yu notes, there is a precarious balance in *Moby-Dick* (125), one which Melville himself describes the same year he published his novel in his essay "Hawthorne and His Mosses":[32] "in certain moods, no man can weigh this world, without throwing in something, somehow like Original Sin, to strike the uneven balance" (*The Piazza Tales* 243). As Hurh has it, Melville's passage suggests that even balance, in the feeling or process of thought, is balanced with imbalance:

> It is in the book Melville was composing at the time he wrote the uneven balance passage – *Moby-Dick* – that he raises the question of thought as a terrifying balance to the level of theme. . . . Two of the book's prominent features, its philosophical stakes and its figurative trope of balancing doubles, are illuminated by a third: the affect of terror that comprehends

> balance by adopting an imbalanced perspective. The paradoxical mode in which a balance can be both balanced and imbalanced at the same time – that the balance itself is weighed against imbalance – is not a logically coherent idea, Melville seems to have concluded, but it is arguably a feeling and, as such, does not itself have to obey logic's strictures.
>
> *(167)*

Balance is not only a problem or outcome of measurement, but of thought – a process of sensation and movement, of placement and displacement, fundamental to thought. And so, it is suggestive that our narrator, who prides himself on being balanced, easy and peaceful, "snug" and "safe" (14), tells us he has "turned into a pillar of salt." Balancing his ease with curiosity, safety with danger, virtue with "sin," or mirroring Bartleby's balanced imbalance with his own imbalanced balance?

As many have noticed, "Bartleby" is a work in two parts (e.g., see John Seelye's *Melville: The Ironic Diagram*, as well as Richard Kopley's "The Circle and its Center in 'Bartleby, the Scrivener'" and Milton R. Stern's *Towards "Bartleby, the Scrivener"*). But do these parts strike the "uneven balance?"[33] Certainly, in its duality "Bartleby" affirms the strange asymmetry of ecogeological and social worlds. And Bartleby himself counters the delicate balance in the office before his arrival, the operational understanding between our narrator and his two, somewhat opposed employees (of temporally opposed humors). The equilibrium he upsets is a caricature of the illusory balance of capitalist imbalance, the inequality of use and surplus value in which Turkey and Nippers are arranged against each other into a symmetry that makes them, in the eyes of the Lawyer, like furniture. Indeed, John H. Randall III foregrounds Turkey and Nippers's view of "themselves [as] machines" as integral to the story's critique "of the inhumanity of the commercial ethos" of New York in the 1850s (141, 144). ("Bartleby's" Dickensian inflection has been noted by several critics, for me the most significant detail of which is the echo of Nemo, the strange and solitary scrivener of *Bleak House*).[34] Rather fittingly, Bartleby's stone presence is a stormy disruption of the business of Wall Street, and of the narrator's carefully circumscribed, "balanced" life. In "The Bartleby Industry and Bartleby's Idleness," Andrew Knighton reads the Lawyer's ultimate failure in terms of a misguided pursuit or practice of balance: "He balances his desire to flee Bartleby with a longing to stay near him; he equivocates between hostility and sympathy; he hastens to dismiss the problem of Bartleby just as he postpones actually dismissing Bartleby himself" (193).

Substance of Bartleby

As McCall argues, "Bartleby" poses the problem of criticism full stop; critics seeking to "prove" their readings seem only to reproduce the copies Bartleby prefers not to (14). What, then, is the substance of "Bartleby"? Is it, as Occupy might have phrased it for a tee shirt, Bartleby's *sub*stance, his stand from below, from which the substrate of Wall Street might rise up? Or is it the problems and possibilities of analogy, as indicative of the practice of reading itself, as Castronovo implies in his close reading of Occupy's reading of "Bartleby"? The lawyer asks himself over and over, however disingenuous it becomes, what's the matter with Bartleby, and we ask ourselves, over and over, what's the matter *of* "Bartleby."

The question beneath all that preoccupies close readers of the text is the one that may (or may not!) occupy Bartleby: the relation between thought and world. More narrowly, in this essay, the relation between (close) reading and an ethics responsible to a damaged (and damaging) world. Bartleby, like Candide a century before him, is a metaphor and no mere metaphor, a man upon whose signification and silence the narrator and critics write, an idea of nature as "the raw material of culture."[35] As we read the narrator search for meaning, construct a discourse of balance and imbalance, we too search for meaning. It is, I have come to believe, a pursuit that always relies on a particular notion of substance, as Aristotle in his *Metaphysics*, and then Adorno in his, had it: *a thing upon which nothing else relies*. In Adorno's analysis, he states that Aristotle's concept of the substantiality of the particular, as that which isn't "predicated on some other, underlying thing . . . that which needs nothing else in order to exist," is "the fundamental thesis of western metaphysics" (28). Adorno points out that Aristotle's work – that is, materialism – and western philosophy generally attempt to rescue the validity of the transcendental moment. In other words, this concept of substance, the "bedrock" (28) of western thought, posits the centrality of transcendence in its denial of transcendental forms. However, if we take a step back from this formulation, we see this concept of substance not only reifies transcendence but is *born from it* (much like the Kabbalastic account of creation Adorno references). It denies the fundamental nature of material relations: contingence and interdependence. In other words, this idea of particular things fundamentally, paradoxically, excludes their existence. It is, equally, the fantasy of a place where the "buck" (of thought) may stop, a stasis beneath reality. As "Bartleby's" insistence on relation insists, there is no place at which the chain of ethical responsibility stops; there is only weight, distance, and the politics of their measurement.

As we know, no one is no one, a Nemo, a ghost. Everyone is someone's child. No one is an island (and no island is an "island," either). For good and for ill, the walls do not hold Wall Street in or out. Nothing is predicated on nothing. All is contingent and in complex relation; the world is both agential and real. Craving the real, as Melville's contemporary Thoreau did,[36] does not have to mean fabricating a place where relation stops. As Thoreau put it in Chapter 2 of *Walden*: the real is that "hard bottom, rocks in place, which we can call reality, and say, This is, and no mistake; and then begin, having a point d'appui, below freshet and frost and fire" (80). Read materially, ecogeologically, reality is only what we call the place at which *we* realize that all is relation, even those hard rocks that seem still in space and time.

In the Anthropocene, this unthinkable age, thinking about reality means realizing our own powerfully significant *in*significance; the inhuman is always more than the ground of the human and human meaning-making. An uncanny embodiment of inhuman nature, Bartleby might seem a ghost to the narrator, a supernatural apparition, but he is as substantial as the earth on which he finally rests. Like the Earth, he is both vulnerable to and beyond human action. And, like the Earth, he is just "there," an appearance without explanation, matter that evolves, and invites endless speculation, just as the text resists ethical closure. The Lawyer's insufficient response to Bartleby and his incomplete biography, the fissure between his rationalizations and unreliability and Bartleby's philosophical and constitutional unreason, provokes our empathy and invites our close reading, as response and *responsibility* (to borrow Derrida's language). Ethics is the knowledge of what we feel, the knowledge that we are all responsible to each other; navigating and negotiating the distances and difficulties of this responsibility is what we call politics.

Without falling into a facile reading of a text as a prescription for philosophical or ecological ills, literature may still be, as Kafka had it, "the axe for the frozen sea inside us" (16). The inhospitality of "Bartleby, the Scrivener: A Story of Wall Street" is a dramatization of the corporeality of the geologic and the geology of embodiment[37] that only becomes more resonant as the present becomes more vulnerable to and beyond human action. Reading hospitality beyond Derrida's epiphany of the embeddedness of the inside in the outside, past our discomfort of "reading people as things" (*Ecocriticism* 103), to a politics of being as belonging in ecogeological terms, "Bartleby" makes a larger place for the inhuman inside us. If we can "stay," as Yusoff has it,[38] or *tarry* with the idea of Anthropocene, close reading itself becomes a means and an end. It is an insistence on knowledge's entanglement with ethics and a reminder of the politics of measurement.

Notes

1 A literalization of Protagoras's claim.
2 "So uniquely vulnerable are the young that the World Health Organization estimates that children younger than 5 bear more than 40 percent of the global burden of disease caused by environmental risk factors and 88 percent of the disease burden caused by climate change." Frederica Perera, "The Womb Is No Protection From Toxic Chemicals," *New York Times*, June 1, 2017. www.nytimes.com/2017/06/01/opinion/toxic-chemicals-pregnancy-fetus.html
3 For example, see www.pbs.org/newshour/show/the-dangerous-global-flood-of-misinformation-surrounding-covid-19 and www.washingtonpost.com/health/2020/02/10/coronavirus-is-spreading-rapidly-so-is-misinformation-about-it/
4 "Bartleby, the Scrivener" was first published in *Putnam's Magazine* in two parts, in November and December 1853, and collected in *The Piazza Tales* with "The Encantadas, or Enchanted Isles" and other stories in 1856. "The Encantadas" was first published in three parts in *Putnam's*, in March, April, and May 1854.
5 See the Introduction to this volume.
6 We surely are "capricious accelerators," though we seem far more rapacious than capricious, and viruses (for just one counter example) mutate and evolve on a scale that

makes the human seem like Aesop's tortoise (though this too is a complex interrelation: see, e.g., Nicholas R. Meyerson and Sara L. Sawyer's "Two-stepping through time: mammals and viruses" in *Trends Microbiology*. 19(6) 2011: 286–94). To return to "The Encantadas, or Enchanted Isles," Melville chose the tortoise as emblematic of "uninhabitableness" (127) and dogged, even suicidal, determination (more on this to follow).

7 In conversation, 2016.

8 See Stacy Alaimo's *Exposed: Environmental Politics and Pleasures in Posthuman Times,* 7, 145.

9 There are too many readings of Bartleby as psychotic or neurotic to list here.

10 Henry A. Murray even uses the term "Bartleby complex" for infantile autism (noted in William P. Sullivan's "Bartleby and Infantile Autism: A Naturalistic Explanation").

11 See Gillian Brown's wonderfully written "The Empire of Agoraphobia."

12 A victim "of both physical and informational entropy who turns himself into a perfect 'isolated system,'" claims Peter Smith in "Entropy in Melville's "Bartleby the Scrivener" (155).

13 Here, Bartleby is a figure of negation that the narrator tries to diffuse, at the end, with "a generalized lament so anodyne that it might apply to anyone, which is precisely the point" (Castronovo 269). From this perspective, one could argue Bartleby is a figure for politics, and the Lawyer for apolitical humanism.

14 A text in two persons and two parts. Melville was fond of diptychs, from the pairs of portraits in *Moby-Dick* to companion pieces, such as "The Paradise of Bachelors" and "Tartarus of Maids."

15 See the article on the Dutch "purchase" of Manhattan on the National Museum of the American Indian website ("America's First Urban Myth?", posted by Dennis Zotigh on August 3, 2011).

16 See Bruce H. Franklin's "The Islands World of Darwin and Melville" in *The Centennial Review* and William Howarth's "Earth Islands: Darwin and Melville in the Galapagos" in *The Iowa Review.*

17 See Hester Blum's "Melville and Oceanic Studies," which argues "Melville's work exemplifies oceanic ways of being" (24). On Melville and ocean ecology, see Elizabeth Schultz's "Melville's Environmental Vision in *Moby-Dick*" in *Interdisciplinary Studies in Literature and Environment.*

18 According to Paul Deane, "things are not inevitably what they seem. On one of the isles Melville finds an exceptionally advanced culture, which he ascribes to the work of pirates" (13).

19 From Norbert Wiener's "The Human Use of Human Beings" (quoted in Peter A. Smith "Entropy in Melville's "Bartleby" 156).

20 In "The Corpse in the Office: Mortality, Mutability and Salvation in 'Bartleby, the Scrivener'" Ronald Wesley Hoag argues Bartleby is "a functional memento mori . . . a figurative corpse" (119).

21 See Amitav Ghosh's *The Great Derangement: Climate Change and the Unthinkable* (2016).

22 Or, as Paul Hurh argues, a performance of "the perfect Cartesian subject" (Hurh 221). For two close readings of philosophers reading "Bartleby," see Kevin Attell's "Language and Labor: Silence and Stasis" and Rachel Cole's "The Lawyer's Tale: Preference, Responsibility, and Personhood in Melville's 'Story of Wall-street.'"

23 The *OED* defines the term as an "artificial reservoir for the storage of water" and notes its first appearance in the *Bible* in 1382 in this apt passage: "Throw ȝe him into the olde sisterne, that is, in wildernes." It is also applied to "various large vessels for water or liquor" (1577), "a pond, or a natural reservoir or depression containing water" (1616), and "a cavity, or vessel in an organism" (1615).

24 *Thrice*: reminding of an opposite expanding enclosure, Coleridge's pleasure dome of Xanadu where water and rock become Earth's respiration (from "Alph, the sacred river, ran/Through caverns measureless to man/ Down to a sunless sea" to "this earth in fast thick pants were breathing" c. 1797/p. 1816).

25 The Lawyer believes at one point that Bartleby refuses because his eyes have been damaged by work. This unfounded belief serves to foreground the narrator's own faulty vision.

26 "Gypsum is found in nature in mineral and rock form. . . . As a rock, gypsum is a sedimentary rock, typically found in thick beds or layers. It forms in lagoons where ocean waters high in calcium and sulfate content can slowly evaporate and be regularly replenished with new sources of water. The result is the accumulation of large beds of sedimentary gypsum" (https://mineralseducationcoalition.org/minerals-database/gypsum/).

27 As Jeffrey Jerome Cohen argues in *Stone: An Ecology of the Inhuman* and elsewhere: "agency is distributed among multifarious relations . . . only in admitting that the inhuman is not ours to control, possesses desire and even will" can we render "human centrality a problem rather than a starting point" (*Prismatic Ecologies* xxiv).

28 See page 22, and then 23, for our Lawyer's need for a "sweet morsel" to satisfy his conscience.

29 As Jamie Lorimer has it, common humanity "masks differential human responsibilities for exposures to planetary change" (quoted in Alaimo 153).

30 See "Loving the Alien: Bartleby and the Powers of Non-Preference" (416–18).

31 "Only Bartleby faces the stark problem of perception presented by the walls. For him external reality thus takes on some of the character it had for Ishmael, who knew that color did not reside in objects, and therefore saw beyond the deceptive whiteness of the whale to 'a colorless, all-color of atheism'" (Marx 89).

32 Appearing in two parts in August 1851 in Duyckinck's *Literary World* as "Hawthorne and His Mosses, By a Virginian Spending July in Vermont." *Moby-Dick* (entitled *The Whale*) was published first in Great Britain in three volumes in October 1851; the first American edition was published November 1851.

33 While Bruce Harvey sees Melville as an author of balance generally (2), Kevin Attell argues that "Bartleby," specifically, is balanced between potentiality and impotentiality (216).

34 *Bleak House* first appeared in installments, from March 1852 through September 1853, prior to the first publication of "Bartleby" in November and December 1853. As McCall mentions, "When he wrote 'Bartleby, the Scrivener' he was just entering his thirty-fourth year. Naturally he turned to Dickens, whose *Bleak House* he had just finished" (30). There are, of course, many readings of the story as an indictment of mid-century capitalism. The best of these in recent years, David Kuebrich's "Melville's Doctrine of Assumptions: The Hidden Ideology of Capitalist Production in 'Bartleby'" and Gillian Brown's "The Empire of Agoraphobia," discuss the story as a self-conscious critique of the narrator's compound ideology and a dramatization of (Bartleby's) justified fear of the marketplace, respectively.

35 Haraway's phrase in *Primate Visions* (13).

36 The very word "real" encompasses the promise of thought on which all thought relies; as the old radical intones in the same section, "Be it life or death, we crave only reality." As Jonathan Culler argues in *On Deconstruction*, in *Walden* the hard bottom, or *point d'appui*, importantly shifts, combines contradictory meanings: "The bottom one can see is too shallow. The figure of the sky as pond combines the desire for a bottom with the depth of bottomlessness. The blackness of the sky is the best natural bottom" (232).

37 "Yet in a century of biology, there is a missing language for geological life as corporeal rather than planetary; as constitutive of subjectivity as well as worlds. . . . The cultivation of this geologic sensibility is an aesthetic ethico-political task" ("Anthropogenesis . . ." 22).

38 Staying with the idea is a process of close reading, an engaged oscillation, as here: "I want to stay with the 'promise' of the Anthropocene as the configuring of an epochal moment of planetary thought, despite, and possibly because of, all the explicit problems that make the Anthropocene both an easy target and a malleable term. These are the problematics associated with grandiose nature-culture divisions, the gendering of 'man' as the subject of this epoch, the imperial basis of western science, climate racism, uncritical social figures such as 'humanity', recourse to the 'population problem', and the mimicking of

global geopolitics in assumptions about reach, scale and control. . . . [A]t the very least, humanity-as-strata forms an ontological rupture with humanist perspectives of humanity as exclusively unified by social forms. So, it is a rebirth in so much as definitions of being must now acknowledge an eternal but shifting mineralogical root; no one is not compromised or enriched by fossil fuels extraction, and in this sense it is a unity from below, but one that is highly differentiated and shot through with relations of power. So staying with the promise of the Anthropocene is rather precarious because it entails taking up the (inhumanist) space that is opened by the concept while refuting the basic architectures of thought that structure that space (where humanity is used as a term of erasure of material and political forms of differentiation)" (Anthropogenesis 8, 9).

Works cited

Adorno, Theodor W. *Metaphysics*. Ed. Rolf Tiedemann. Trans. Edmund Jephcott. Stanford UP, 2000.

Agamben, Giorgio. *Potentialities: Collected Essays in Philosophy*. Trans. Daniel Heller-Roazen. Stanford UP, 1999.

Alaimo, Stacy. *Exposed: Environmental Politics and Pleasures in Posthuman Times*. Minnesota UP, 2016.

Attell, Kevin. "Language and Labor, Silence and Stasis: Bartleby Among the Philosophers." *A Political Companion to Herman Melville*. Ed. Jason Frank. UP of Kentucky, 2013: 194–228.

Blanchot, Maurice. *The Work of Fire*. Trans. Charlotte Mandell. Stanford UP, 1995.

———. *The Writing of Disaster*. Trans. Ann Smock. U of Nebraska P, 1986.

Blum, Hester. "Melville and Oceanic Studies." *The New Cambridge Companion to Melville*. Ed. Robert S. Levine. Cambridge UP, 2014: 22–36.

Brown, Gillian. "The Empire of Agoraphobia." *Representations*. 20. Autumn 1987: 134–57.

Castronovo, Russ. "Occupy Bartleby." *J19: The Journal of Nineteenth-Century Americanists*. 2 (2). 2014: 253–72.

Clark, Nigel. *Inhuman Nature: Sociable Life on a Dynamic Planet*. Sage, 2010.

Clark, Timothy. *Ecocriticism on the Edge: The Anthropocene as a Threshold Concept*. Bloomsbury, 2015.

Cohen, Jeffrey Jerome. "Introduction." *Prismatic Ecologies: Ecotheory beyond the Green*. Ed. Cohen. Minnesota UP, 2013.

———. *Stone: An Ecology of the Inhuman*. Minnesota UP, 2015.

Cole, Rachel. "The Lawyer's Tale: Preference, Responsibility, Personhood in Melville's 'Story of Wall-Street'." *Melville's Philosophies*. Eds. Branka Arsić and K.L. Evans. Bloomsbury, 2017.

Culler, Jonathan. *On Deconstruction: Theory and Criticism after Structuralism*. Cornell UP, 1982.

Deane, Paul. "Herman Melville: The Quality of Balance." *Serif*. 7 (2). 1970: 12–17.

Deleuze, Gilles. *Essays Critical and Clinical*. Trans. Daniel W. Smith and Michael A. Greco. Minnesota UP, 1997.

Derrida, Jacques. *The Gift of Death and Literature in Secret*. Trans. David Wills. Chicago UP, 1995, 2008.

———. *On Hospitality: Anne Dufourmantelle Invites Jacques Derrida to Respond*. Trans. Rachel Bowlby. Stanford UP, 2000.

Fisher, Marvin. "'Bartlebym' Melville's Circumscribed Scrivener." *Southern Review*. 10. 1974: 59–79.

Franklin, Bruce H. "The Islands World of Darwin and Melville." *The Centennial Review*. 11 (3). 1967: 353–70.

Ghosh, Amitav. *The Great Derangement: Climate Change and the Unthinkable*. Chicago UP, 2016.

Gullestad, Anders. "Loving the Alien: Bartleby and the Powers of Non-Preference." *Text, Action and Space: Exploring Textual Action*. Ed. Lars Saetre, Lars, Anders Kristian Strand, Mads Thygesen. Aarhus University Press. 2010: 395–422.

Haraway, Donna. "Anthropocene, Capitalocene, Chtulucene: Staying with the Trouble, 5/9/14." https://vimeo.com/97663518.

———. *Primate Visions: Gender, Race, and Nature in the World of Modern Science*. Routledge, 1990.

Hardt, Michael and Antonio Negri. *Empire*. Harvard UP, 2000.

Harvey, Bruce. "Science and the Earth." *A Companion to Herman Melville*. Ed. Wyn Kelley. Blackwell, 2006: 71–82.

Hoag, Ronald Wesley. "The Corpse in the Office: Mortality, Mutability and Salvation in 'Bartleby, the Scrivener'." *ESQ: A Journal of the American Renaissance*. 38 (2). 1992: 119–42.

Howarth, William. "Earth Islands: Darwin and Melville in the Galapagos." *The Iowa Review* 30 (3) 2000/2001: 95–113.

Hurh, Paul. *American Terror: the Feeling of Thinking in Edwards, Poe, and Melville*. Stanford UP, 2015.

Kafka, Franz. *Letters to Friends, Family and Editors*. Trans. Richard Winston and Clara Winston. Schocken, 1990.

Knighton, Andrew. "The Bartleby Industry and Bartleby's Idleness." *ESQ: A Journal of the American Renaissance*. 53 (2). 2007: 184–215.

Kopley, Richard. "The Circle and its Center in 'Bartleby, the Scrivener'." *ATQ*, 2 (3). 1988: 191–206.

Kuebrich, David. "Melville's Doctrine of Assumptions: The Hidden Ideology of Capitalist Production in 'Bartleby'." *The New England Quarterly*. 69 (3). Sept. 1996: 381–405.

Marr, Timothy. "Melville's Planetary Compass." *The New Cambridge Companion to Melville*. Ed. Robert S. Levine. Cambridge UP, 2014. 187–201.

Marx, Leo. "Meville's Parable of Walls." *Bartleby the Inscrutable*. Ed. M. Thomas Inge. Archon, 1979.

Mordecai, Marcus. "Melville's Bartleby as a Psychological Double." *Bartleby the Inscrutable*. Ed. M. Thomas Inge. Archon, 1979.

McCall, Dan. *The Silence of Bartleby*. Cornell UP, 1989.

Melville, Herman. *The Writings of Herman Melville volume 9. The Piazza Tales and Other Prose Pieces 1839–1860*. Ed. Harrison Hayford et al. Northwestern UP, 1987.

Randall, John H. III. "Bartleby *vs*. Wall Street: New York in the 1850s." *Bulletin of the New York Public Library*. 78. 1975: 138–44.

Rogan, Michael Paul. "Melville and the Slavery of the North." *Modern Critical Interpretations: Billy Bud, Benito Cerano, Bartleby the Scrivener*. Ed. Harold Bloom. Chelsea House, 1987: 107–26.

Schultz, Elizabeth. "Melville's Environmental Vision in *Moby-Dick*." *Interdisciplinary Studies in Literature and Environment*. 7 (1). 2000: 97–113.

Seelye, John. *Melville: The Ironic Diagram*. Northwestern UP, 1970.

Smith, Peter A. "Entropy in Melville's 'Bartleby the Scrivener'." *The Centennial Review*. 32 (2). 1988: 155–62.

Stern, Milton R. *Towards 'Bartleby the Scrivener'*. U of Toronto P, 1979.

Sullivan, William P. "Bartleby and Infantile Autism: A Naturalistic Explanation." *The Bulletin of the West Virginia Association of College English Teachers*. 3 (2). 1976: 43–60.
Thoreau, Henry D. *Henry David Thoreau: Walden*. Library of America, 1991.
Waugh, Charles. "We Are Not a Nation, So Much as a World: Melville's Global Consciousness." *Studies in American Fiction*. 33 (2). 2005: 203–28.
Yu, Beongcheon. "Ishmael's Equal Eye: The Source of Balance in Moby-Dick." *ELH*. 32 (1). 1965: 110–25.
Yusoff, Kathryn. "Anthropogenesis: Origins and Endings in the Anthropocene." *Theory, Culture & Society*. 33 (2). 2016: 3–28.
———. *A Billion Black Anthropocene or None*. Minnesota UP, 2019.
Žižek, Slavoj. "Notes towards a Politics of Bartleby." *Comparative American Studies*. 4 (4). 2006: 375–94.
———. *The Parallax View*. MIT Press, 2006.
Zotigh, Dennis. "American's First Urban Myth?" *The National Museum of the American Indian*. August 3, 2011. https://blog.nmai.si.edu/main/2011/08/americas-first-urban-myth.html

5

CLOSE READING AT THE END OF TIME

Mark C. Long

While many, if not most, literary critics eschew the assumptions of New Criticism, Jonathan Culler reminds us that even for the New Critics, the work of close reading was not "primarily to resolve difficulties but above all to describe them, to elucidate their source and implications" ("Closeness" 22). A general term, as Peter Middleton puts it, "for a heterogeneous and largely unorganized set of practices and assumptions" (5), close reading has nevertheless become the name for an accepted interpretive framework. These commonplace interpretive assumptions and practices are disclosed in the ways teachers and scholars talk about reading, as well as processes of literary and cultural transmission; and indeed hermeneutics (in contrast to poetics) is privileged in a range of instructional settings – most notably as the preferred method for teaching reading in secondary or postsecondary institutions in which students generate "readings" of texts to demonstrate evidence of competency, critical literacy, and literary sophistication.

This essay will describe this framework of assumptions and practices to consider some of the ways close reading is debated in recent critical studies concerned with past and present ways of thinking about nature. Of particular interest will be critical projects preoccupied with scales of time – for instance, Jonathan Sachs' concern with "temporal dissonance" (317) in Romantic writing, and the "paradoxical temporality" David Farrier explores in Anthropocene poetics (6) – or what I am calling "close reading at the end of time."[1] For these proposals for reading in the Anthropocene make visible the strengths and the limits of close reading and, as a consequence, the institutional fragility of disciplinary formations in the humanities. These inextricable challenges suggest that the future of close reading depends on deeper questioning of the protocols for scholarly reading and the hermetic institutional structures that reinforce and reproduce particular ways of reading. The promise of close reading in the Anthropocene,

I will argue, begins in first accepting the limited social value of demonstrative reading and then embracing the constitutive power of close reading as a collective practice.

Debates over close reading have for the most part focused on the legacy and coherence of literary studies. Most recently, these conversations about reading – and the critical, philosophical, and institutional concerns that arise in the study of reading practices – have emerged in proposals for expanding humanist practices in the environmental humanities.[2] Tobias Menely and Jesse Oak Taylor explain in *Anthropocene Reading*, for example, that the multidisciplinary problem of the Anthropocene has renewed conversations about "the legacy of theory, the future of method, and the coherence of literature as an object of study" (10).[3] They link the "pronounced methodological disquiet" in twentieth-century literary studies to the "crisis" in the discipline that has led to questions about "how we justify the resources dedicated to our work, reading, teaching, and writing about literature in an age of neoliberal austerity and STEM ascendance." They trace ecocriticism's engagements with energy and matter and recourse to scientific principles, "first biology and ecology, then sociology and political science, and now stratigraphy and Earth system science" (12). They call attention to new materialists and posthumanists who broaden "our conception of signs, agents, and relations so as to resituate humans, and human-meaning making, in a broader constellation of beings" (11). Serpil Opperman, to take one example, elaborates how material criticism attempts to rethink the conceptual geographies of the Anthropocene by seeing "the world in terms of the narrative potentials that inhere in all agencies and understand that humans are not the only storied beings in this earthly life" (11). Opperman argues that to counteract the disorienting framing of the Anthropocene, "material ecocriticism reads the world through narrative agencies that are thick with material-discursive archives of survivals and extinctions, always confounding the human." For Menely and Taylor, the project of the Anthropocene suggests a model for reading "flows, trajectories, and systems that go beyond national borders and human time scales, while at the same time attending to the interplay of these systemic relations through fine-grained analysis."

These representative proposals to read across natural and cultural histories are grounded in the questions Rob Nixon helped articulate for literary and cultural historians in an age of environmental crisis. "How can we convert into image and narrative the disasters that are slow moving and long in the making, disasters that are anonymous and star nobody, disasters that are attritional and of indifferent interest to the sensation-driven technologies of our image-world? How can we turn the slow emergencies of slow violence into stories dramatic enough to rouse public sentiment and warrant political intervention, these emergencies whose repercussions have given rise to some of the most critical challenges of our time?" (3). These questions, it is important to add, define literary and cultural criticism as a form of cultural work that should "rouse public sentiment" and "warrant political intervention." Among the challenges for these projects has been that arguments for the purposes and value of reading and writing infrequently reach beyond a narrow audience of people preoccupied with problems of literary and cultural representation. The result, as John Guillory explains in "The Way We Read," is that the sites, experiences, and conventions of lay and professional reading are so different that "a certain mutual misunderstanding has arisen between lay and professional readers, a sense on the one side that the professors have betrayed their traditional duty to offer guidance to lay readers, and on the other, that sophisticated and politically vigilant techniques of reading can have transformative effects well beyond the sphere of their practice" (33).[4]

Proposals to expand humanist practice in the Anthropocene have at the same time found it difficult to disengage close reading from the problem of literary form. Nicholas Gaskill describes the problem as "measuring the distance between twentieth- and twenty-first century formalisms" to articulate what, "in our moment of methodological inventiveness, close reading might become" (520). For Menely and Taylor, the Anthropocene is an opportunity "for literary studies to test and transform its methods by examining how a symbolic domain might, or might not, index a historicity that exceeds the human social relation and encompasses planetary flows of energy and matter" (5). But for Jeffrey Cohen, it may turn out that the "interpretive practices derived from a linguistic archive" will not be of much use in coming to terms with the Anthropocene.[5] Indeed a move toward what Nicole Merola calls "phenomenon-focused" approaches raises questions about "what literary scholarship might look like if directed toward nonanthropogenic forms of writing, biogeochemical traces neither created by nor meant for human readers" (823).[6] It is worth asking, then, whether the commitments entailed in these emergent ways of reading – and the theoretical traditions that inform these reading practices – suggest the contours of a collective project. For the temporal and geographic scales of these phenomena-based approaches may turn on the possibilities and limits of reading across scales of geological or "deep" time.

In other words, the Anthropocene forces a reckoning with the strengths and weaknesses of the inherited tradition of a discipline organized around the study of literature.[7] Although these examples of reading, it turns out, are repurposing interpretive vocabularies to describe conceptual scales of time (close and distant) and proximity (surface and depth). And so to better understand what close reading might be we need to clarify what we hope to gain from close reading. Most proposals for "reading the world," to use Opperman's formulation, may in fact suggest a need to articulate a way of reading that is neither close nor deep. That is, the "flows" and "trajectories" and "systems" that offer interpretive freedom to move across domains come at a cost: the loss of specificity. This loss is precisely the form of attention associated with a particular medium (literature) and a form of inquiry (literary study). Moreover, as Hayles reminds us, the mode of deep attention associated with close reading was developed to address complex problems in a single medium. Reading preoccupied with the Anthropocene may therefore call for hyper-reading, a mode that in Hayles' description "excels at negotiating rapidly changing environments in which multi foci compete for attention" (188).[8] Menely and Taylor propose that we "read the Anthropocene as a literary object and at the same time to recognize the Anthropocene as a geohistorical event that may unsettle our inherited practices of reading" (5). It may be more likely, then, that what is unsettled is the disciplinary work organized around the cognitive mode of deep engagement with objects such as literature.

This trajectory for critical practice therefore raises not only disciplinary but institutional questions. On the one hand, we have Menely and Taylor claiming that "in the Anthropocene, all scholars are called upon to become Earth system humanists, which involves thinking about how these systems interrelate with, internalize, and destabilize one another" (4). On the other hand, we have Timothy Clark contending, in "Derangements of Scale," that coming to terms with the relationship between climate change and literary and cultural criticism "will not be a matter of inventing some new method of reading per se, for its most prominent effect is of a derangement of scales that is also an implosion of intellectual competences." For Clark, "It is far easier for critics to stay inside the professionally familiar circle of cultural representations, ideas, ideals and prejudices, than to engage with long-term relations of physical cause and effect, or the environmental costs of an infrastructure." These kinds of questions, we would do well to remember, "involve nonhuman agency and which engage modes of expertise that may lie outside the humanities as currently constituted." It may follow that the challenge of scholarship in the Anthropocene may be the disciplinary limits of professional training as well as the personal investments in humanist identities and ideological commitments. In brief, as Clark explains the challenge, "the humanities as currently constituted make up forms of ideological containment that now need to change." This provocation suggests, at the same time, that an "Earth system humanism" that seeks to "resituate humans, and human-meaning making, in a broader constellation of beings," may be a modest proposal indeed. For the recent history of criticism and theory remind us that disciplinary structures and modes of inquiry are not necessarily disposed to ideological or organizational change.[9]

Any (re)turn to the practice of close reading in the Anthropocene, it follows, must begin with an openness to questioning our justifications for reading and writing – especially the ways we associate our scholarly and pedagogical modes of reading with actual effects in the world. For no matter how earnest our convictions are about the transformative potential of our intellectual work there is little more than anecdotal evidence for "the actual effects in the world of what we do, either as teachers or as scholars" (*Ethical Practice* 31). With no credible way to assess these effects, "and in the absence of such determinable measures," Guillory argues, "we tend to indulge" in what he calls "the fantasy of transforming the world to a degree vastly greater than can reasonably be expected of perhaps any disciplinary practice."[10] These unsettling questions about the consequences of literary and cultural work might even make it reasonable to conclude that the familiar routine of scholarly demonstration and exchange in professional communities holds little promise for the future of close reading. Guillory's critique, to be clear, is that if "any and every effect that we have in the world must be achieved through a practice of reading" then the "limits of this reading practice constitute the limits of disciplinary practice." What follows is what I wish to propose in the remainder of this essay: that the reading we do beyond the parochial concerns of scholarly communities, that we enact in proximity to others, may be the closeness of reading that matters most.

This practice of reading in proximity to others is a socially situated, radically contingent, and potentially more consequential activity than the disciplinary routine of circulating scholarly or professional "readings" that advance a particular theory or a demonstration of its claims. To effectively refocus a professionalized preoccupation with scholarly reading on teaching and learning, however, requires rethinking – and actually resisting – the entrenched and hierarchical incentives and rewards that devalue pedagogical labor. The pedagogical trajectory I envision for close reading must at the same time reimagine classroom work – that is, the reading and thinking and writing of students – as a collective scholarly project. What might close reading become if the practice were reconceived as a fundamentally collective project within the everyday collaboration of teachers and students? At the very least, close reading would offer a collaborative response to the climatic regime we have come to call the Anthropocene: for the attention of readers cultivated in the practice of close reading together would engage more people with the cultural problem of attention to the particular urgencies of living in our fragile and precarious world.

This collaborative response would begin with our students. For although climate change is affecting everyone alive today, as Sarah Jacquette Ray has recently argued, our students are the "climate generation" who have not only lived their entire life with the effects of climate change, but "who are poised to organize and bring about real change" (3).[11] A student-focused model for cultivating resilience and "embracing life" in the Anthropocene calls us to engage more deeply in the experiences we share with our students in the classroom. Close reading, defined as a collective practice, and aligned with the traditional association of close reading with attention, would provide one of the most useful modes of working together in the collective, uncertain, and ongoing activity of living in the world. This future for close reading is, at the same time, a proposal for a more widespread scholarly engagement with contexts of teaching and learning. While this proposal might first appear to devalue the material and social communities of scholarly exchange it is, rather, a question of first priorities. How best do we respond collectively to the institutional precarity and as well as the ethical questions we face in the Anthropocene? My wager is on the primary value of intellectual work in teaching and learning to move us from forceful – but, in the end, mostly parochial – modes of critique to more generous modes of thinking that unfold in the necessary and ongoing work of reading together.[12]

To begin, we need a more sufficient account of reading itself: both a robust critique of how we talk about close reading and a definition of close reading as a meaning-making activity. For when the predication shifts from "reading closely" to "doing a close reading" the conceptual metaphors we are working in not only determine the language to describe close reading but also reinforce widely shared commonplaces about the act (or event) of reading. For instance, teachers and students routinely say that in reading "closely" and "deeply," or reading "between the lines," we discover meaning.[13] These conceptual metaphors are long-standing, and unfortunate, for the interpretive purpose of close reading that reinforces meaning "in" the text is predicated on the conceptual difference between surface and depth and the related idea of meaning hidden in or under words.[14] This discourse persists because spatial metaphors most often function as proxies for the reader's attention to text, a disposition toward an object of study, or an analytical method for integrating the parts of a text into its whole. And as close reading has evolved into a way of "doing" a close reading the complex process of constructing intelligibility reproduces assumptions about the singular value of particular literary objects and becomes increasingly dependent on the presence and "the ethical charisma of the literary translator or messenger" (Love "Close" 374).

We need to understand the pedagogical paradox that arises when the practice of close reading is associated with student agency and empowerment as well. For the association of close reading with empowerment "has a vaguely ethical air of making that reading attentive and careful" even though "it leaves entirely to the discretion (or the unexamined predispositions) of 'close' readers what they attend to or what they make of what they attend to" (Bialostosky 112). The paradox arises when sophisticated readers come face-to-face with a classroom of students with unexamined dispositions and unsupported judgments they will often default to performing a reading of a text. "What students learn from their instructors' expertise (when instructors are true experts) is that instructors can produce persuasive and comprehensive readings or interpretations of texts beings studied and that the students themselves can't do it," Sheridan Blau explains (271). Close reading too often becomes a lesson in teacher dependence as students sense their prior experiences as readers and repertories of discursive knowledge are inadequate to the hermeneutical work at hand.

We need to begin with close reading as a meaning-making activity. This activity starts with the enduring association of close reading with *attention* and the widely shared endorsement of attention as a desirable skill. After all, as Guillory puts it, "the most common and least technical formulation, close reading means paying attention to the words on the page" ("How" 11). Smith, to take another example, notes "the ability and disposition to read texts attentively, one by one (in addition, of course, to digital sophistication) . . . is a practice that we all have a stake in preserving" (73). Or as the reading theorist Louise Rosenblatt explains, reading as a meaning-making activity is an event that unfolds in time and involves a continuum of attention, whether one is reading a poem or a scientific report. But the real problem of placing that continuum at the center of our work with students, according to Rosenblatt, is that in our talk about the complex act of reading we find it difficult "to free ourselves from unscrutinized assumptions implicit in the usual terminology and in the very structure of our language" (16). Rosenblatt's theory of reading, I am suggesting, provides a theoretically sophisticated and practical alternative to current-traditional misapprehensions associated with the idea that meaning is "in a text," or that meaning is simply generated in an individual or community of readers.

Rosenblatt's account of the reading process liberates close reading from either the singular focus on the text, a charge often levelled at the New Critics, or in the reader-response theories that arose to move beyond the textual focus of close reading associated with the New Criticism. More significantly, for the purposes of the brief elaboration I am able to include here, Rosenblatt's transactional approach emphasizes the event of reading as reciprocal and constitutive.[15] Her approach provides a vocabulary in a scholarly tradition of thinking about reading as a threshold event in the ongoing challenge of becoming more aware of ourselves, the language we use, and our embeddedness in the world. Perhaps most significantly, Rosenblatt's transactional paradigm requires a shift in habits of thinking so that the "stimulus-response, subject-object, individual-social dualisms," which serve as defaults can "give way to recognition of transactional relationships" (3). These alternative habits of thinking orient how we think about the individual, in Rosenblatt's words, "as a part of nature, continually in transaction with an environment." Challenging theories of the text or reader or context that fail to account for the transactional process of "event" of reading, Rosenblatt calls on an ecological framework in which human activities and relationships "are seen as transactions in which the individual and social elements fuse with cultural and natural elements." A definition of close reading grounded in Rosenblatt's transactional theory, too, reinforces reading as a transferable skill – a methodology for a pedagogy of literary and cultural studies grounded in a philosophical and anthropological outlook on how we might more productively attend to the world.

Finally, while the attitudes or stances we may take to particular texts may differ, a continuum of reading strategies applies in every case. For it is untenable to treat a text, Rosenblatt insists, either "as an isolated entity or to overemphasize either the author or reader" (27). The future of close reading I am imagining here, then, depends on a sufficient account of literate activity. Assumptions about textual autonomy attributed to New Criticism, and that persist in the practice of close reading, often obscure the basic conditions that make intelligibility possible: that every act of reading begins *in medias res* – "at a particular point in a history of reception," to call on the words of Leroy Searle.[16] "It may be as narrow as our own prior training or experience, or as wide as an institutionalized tradition, integrated with systems of belief and common conceptions that we may or may not have had any occasion to question – and in that sense qualifies for us individually as a 'beginning' in the onset of a particular transaction" (2). This account of reading may appear to be obvious. But it is very often obscured in the hazy presumption that a text or expression in language has meaning. "When we ask what a word 'means,'" Searle explains, "that is already an instance in a history of reception: it is a way to teach someone how a word is *used*, or to illustrate how it is related to other words, in a context of an empirical situation we all share, at least potentially" (2–3). It is precisely because theories of language and society are always already in play that our students need to recognize their preconceptions through experiences of close reading, and to experience the actual situation we find when thinking with others.

In the precious but precarious institutions that make possible traditions of inquiry students experience how we come to know the world, they are introduced to ways of knowing that world, and they learn that the genesis of knowledge is an ongoing event. The practice of close reading becomes a catalyst for experiencing the event when the empirical skill of description leads students to the awareness that any text is embedded in any number of nonliterary things. An awareness of these entanglements frees students to understand that individuals and communities are necessary but wholly insufficient to the social and cultural work of interpretation. Such entanglements, moreover, lead students to recognize what Bruno Latour describes as distributed agency. "A work of art *engages* us," he writes, "and if it is quite true that it has to be interpreted, at no point do we have the feeling that we are free to do 'whatever we want' with it. If the work needs a subjective interpretation, it is in a very special sense of the adjective: we are subject to it, or rather we win our subjectivity through it" (*Inquiry* 241). Latour's emphasis on the event or transaction of reading is not limited to esthetics or to literary objects because both literary objects and esthetics are not limited to any single domain.

Latour's emphasis leads the close reader into the networks in play in the event of reading. For Latour, "The great advantage of this mode of understanding is that it allows the analyst as much freedom as that of the actors in the weaving of their worlds; it frees the field entirely from its organization into domains." This is especially true "when we learn to liberate ourselves from some of the supposedly uncrossable borders . . . between nature and culture, for example, or power and reason, the human and the nonhuman, the abstract and the concrete" (62). The activity of tracing networks requires the attention of close reading at the same time it situates the text in the networks of production and reception that contribute to its constitutive presence in the ongoing event of reading. To define close reading as an event may be closer to the anthropological mode Latour elaborates than we might think. For to trace a textual network, much like the anthropologist in search of an essence of a situation, is "always to reconstitute by a trial (an investigation is a trial, but so is an innovation, and so is a crisis) the antecedents and the consequences, the precursors and the heirs, the ins and outs, as it were, of a being" (41). The trajectories that constitute the same general form as a network, finally, offer a vocabulary for approximating the free play of the imagination, "defined by leaps, discontinuities, hiatuses" (41), in the work of meaning-making that is close reading.

Attention to the lived experience of a reader in a world of relations suggest that close reading would necessarily explore more fully the depths of attachment that, according to Rita Felski, fall outside the prevailing ways of thinking in literary studies. "Interpretation, we might say, constitutes one powerful mode of attachment." Although within a Latourian framework, she asserts, "we do not probe below the surface of a text to retrieve disavowed or repressed meanings, nor do we stand back from a text to 'denaturalize' it and expose its social constructedness." As a result, "the distinction between a knowing critic and an unknowing text – or a naive reader – crumbles away. Reading becomes a matter of composing and cocreating, of forging links between things that were previously unconnected" ("Latour" 741).[17] Once again, we sense the literary work of close reading as more than an event of the text and the reader. It is rather a more complex attending to the event of reading as situating the reader – in a community, that is, beyond the self. Or as Matthew Crawford argues in *The World Beyond Your Head*, as individuals, "we find ourselves situated in a world that is not of our making and this 'situatedness' is fundamental to what a human being is" (26).

Crawford's work provides a useful elaboration how the environment "constitutes" rather than "compromises" the self. His argument is that attention is at the core of a "constitutive or formative process" (25) in what he calls "the struggle for individual agency at the center of modern life."[18] Understanding the self as situated provides a conceptual framework for positioning close reading as a pedagogical activity that joins us to the world and to others. This understanding is difficult because "the ideal of autonomy" in modern life paradoxically works against "the development and flourishing of any rich ecology of attention – the sort in which minds may become powerful and achieve genuine independence" (25). Crawford argues that "the experience of *attending* to something isn't easily made sense of within the prevailing Western anthropology that takes autonomy as the central human good" (24). His case is that "Enlightenment legacy of autonomy talk" does not align with what we have learned about the social nature of what we know (250–51). Through sustained practice we experience (Crawford's word is "inhabit") what he calls "ecologies of attention" that place us *in* the world" (246).

Crawford, in elaborating attention as both an individual and cultural challenge, raises the question whether it possible to define close reading as an immersive, embodied experience that provides access to new perceptions – of particulars, emergent patterns of relation, and identifications with the world. His account of human agency and creativity, moreover, is predicated on the assumption that "membership in a community is a prerequisite to creativity" (129). Might we work toward a more productive encounter with texts in an attentional commons in which the practice of scholarly reading is replaced by a transactional experience of reading – an event in which students are subject to fewer representations of reading experiences and more experience of their own close reading in a classroom designed around the kind of close attention to language we value? Might the classroom inculcate students in a tradition of close reading that subordinates the student to the master teacher who is aware of the paradox of teaching described above, is prepared to enact a pedagogy of meaning-making between the teacher and the students, and understands that the supposed freedom of the autonomous self is not constrained by the heteronomous world beyond the self?

The argument for reclaiming mental autonomy in an age of distraction, as Crawford suggests, requires changing our cultural understanding of individual agency – a change that the writer Amitov Ghosh describes in *The Great Derangement: Climate change and the Unthinkable*. What makes such a change so difficult, Ghosh explains, is that "at exactly the same time when it has become clear that global warming is in every sense a collective predicament, humanity finds itself in the thrall of a dominant culture in which the idea of the collective has been exiled from politics, economics, and literature alike" (80).[19] The change is all the more challenging because of the organization of academic institutions and the interests of the disciplines. These structures exemplify the dominant culture in the everyday practices of scholarly work – and the activities of reading and writing in the classroom. They reflect the isolation and fragmentation of self-interested pursuits. Teaching philosophies and practices, moreover, are guided by scholarly aspirations for students to develop the sophisticated forms of literacy we value or merely instrumental learning outcomes for the literary activities of reading and writing. Considered together, these standing investments in the literary practices of reading and writing suggest an unsettling truth: that our culture of literacy exemplifies, in the words of Ghosh, "the broader imaginative and cultural failure that lies at the heart of the climate crisis." Developing a pedagogical model of reading to address this failure therefore requires thinking about close reading as a collective activity, to call on Ghosh's words, in a "less provincial way to prepare us to inhabit a world that has become common at last" (*An Inquiry* 292).

But to build a case for close reading as a collective activity will require a further development in our thinking about both the methods and value of scholarly reading. "Academics tend to repress consideration of variety in reading practices due to assumptions that everyone reads (or ought to read) the way we do professionally," Heather Long argues, "privileging the cognitive, ideational, and analytic mode" (110). "The solitary reader," Long explains, "suppresses recognition of the infrastructure of literacy and the social or institutional determinants of what is available to read, what is "worth reading," and how to read it" (110). Just as Linda Brodkey's attention to the image of the solitary writer shaped how we think about the writing process our image of reading as a solitary activity shapes how we think about the event of reading. The trope figures the writer, and by analogy the reader, as "withdrawn from the world and suspended from human community and action" (105). Documenting the iconography that figures reading as a solitary activity, Long turns her focus to "some of the consequences of constructing textual interpretation as a fundamentally solitary practice" and the ways in which these constructions efface "the collective nature of reading" (107). And as I have already suggested, to consider close reading in this way involves thinking more about the contexts in which we close read and the ideologies associated with the academic and cultural authority of teaching reading.[20] The challenge is not so much pedagogical as it is questioning the normative conventions of scholarly reading that are reproduced, in subtle ways, in the classroom – conventions that are rooted, in what Long calls "the hegemony of an associated model of how culture works" and a "trickle down model of cultural dissemination" that assumes that all "innovative ideas and values originate with transcendent high cultural figures and are delivered by abstract processes – and in diluted form – to the lower (and in this model, relatively passive) levels of sociocultural hierarchy" (124). Might reimagining the experience of literary engagement as a collective activity make visible for ourselves, as well as others, some of what makes reading compelling for those readers who do not have access to a group with which to discuss the texts?

Reimagining the experience of literary engagement as a collective activity redefines close reading as a relational activity. Close reading makes available the indices of attention that arise in an experience of literary engagement that includes both the reader, the collective of readers in a classroom setting, and the social processes of attending to reciprocal and mutually constitutive relations. Moreover, this description of close reading helps to generate a more sufficient vocabulary for what is actually happening at the level of the individual reader. "Reading," Rita Felski reminds us, "is a matter of attaching, collating, negotiating, assembling – of forging links between things that were previously unconnected." It is not, "a question of plumbing depths or tracing surfaces." While Feliski's project shapes a discourse of attachment in reading, in contrast to the modes of detachment that predominate in suspicion and critique, her argument for "postcritical reading" draws the discourse of actor-network theory into a productive relation to hermeneutics, so as to "better do justice to the transtemporal liveliness of texts and the coconstitution of texts and readers – without opposing thought to emotion or divorcing intellectual rigor from affective attachment" (173). This vocabulary offers another way to (re)turn to close reading as a collective activity that fosters dialogue, debate, and self-correction that we all know, when successful, leads to ideas better than any one person might produce alone. It is the kind of language we need to repurpose and extend a tradition of transactional and constitutive reading, together, with our students, in a community of practice. The future of close reading is in these ways an ethical response to the challenge of the Anthropocene – a pedagogical wager on the affordances of close reading in a culture of literacy that might offer our students ways of thinking and habits of mind to orient themselves toward one another, to the past, and to the planetary future we share.

Notes

1 This encounter between human time (at its end) and deep time at the same time echoes Bruno Latour's description of the present condition in *An Inquiry into Modes of Existence: An Anthropology of the Moderns*, in which he elaborates "an entirely new situation: behind us, attachments; ahead of us, ever more attachments. . . . End of modernization. End of

story. Time to start over." Latour goes on to say that "'Gaia,' the 'Anthropocene' era, the precise name hardly matters" (10). Also see Lecture Six, "How (not) to put an end to the end of times?" in Latour, *Facing Gaia: Eight Lectures on the New Climatic Regime*, especially 184–219.

2 See Lawrence Buell's earlier overview of these developments, and their relevance to the present discussion of methods of reading, in his account of ecocriticism as "a quest for adequate models of inquiry" (10). For an account of reading in first- and second-wave ecocriticism, see my "Introduction: Ecocriticism as a Practice of Reading" to a special issue of *Reader: Essays in Reader-Oriented Theory, Criticism, and Pedagogy*.

3 The essays in *Anthropocene Reading* offer illustrative attempts to read energy regimes, natural histories, and planetary systems. The tropes of scale in the essays include "close," "slow," "deep," "surface," and "distant." Menely and Taylor point out that with "something like the Anthropocene – multiform, multiscalar, multicausal, multitemporal – a commitment to methodological consistency may be exactly the wrong approach" (13). So they reframe the question of method as a question of approach. The editors make use of the metonymic shift in the logic of close reading that moves from particular features of a text to the meaning or significance of the whole. See Jin, "Problems of Scale in 'Close' and 'Distant' Reading."

4 Also see "The Ethical Practice of Modernity: The Example of Reading," for Guillory's elaboration of the argument that reading practices in the academy have become nearly incommensurable with lay reading practices.

5 See Jeffrey Cohen, "Anarky." However, also see Clark, who argues that proposals to read "phenomena" must account for the "crucial differences between reading a literary text at multiple scales and the function of scales in scientific modeling and explanation. In such modeling, suppression of detail is seen as strength of work at large scales, where broad patterns can emerge overriding individual variations. A literary reading clearly works in no such way."

6 See Merola's "Archives of Ecocatastrophe; or, Vulnerable Reading Practices in the Anthropocene" for an especially illuminating review essay of recent books by Alaimo, Heise, Rigby, and Wallace – four authors who "propose and practice forms of reading meant to help readers confront the Anthropocene" and who demonstrate how "reading for human-nonhuman entanglements might refract, register, and generate new onto-ethico-epistemologies for constructing better presents and futures, those attentive to the needs of both human and nonhuman creatures" (822).

7 For Gaskill, the "New Critical project makes no sense apart from the postulation of an aesthetic dimension of experience" and a method of close reading "forged to register that which makes this dimension intelligible (aesthetic form)" (520). We need to ask, Gaskill insists, "whether and how the New Critical methods can work apart from the New Critical assumptions. Not that we would have to advocate their exact ideas about the aesthetic – but we would likely need *some* notion of the specificity of art as a way of thinking or understanding if a practice like close reading is to be used a means of producing evidence" (521).

8 In addressing concerns among educators who value close or deep attention in relation to brain research in neuroscience, Hayles argues that a case can be made that "hyper attention is more adaptive than deep attention for many situations in contemporary developed societies" (194). Might the assets of hyper attention that may be suited to, say, "the air traffic controller who is watching many screens at once," or perhaps the earth system humanism proposed by Menely and Taylor? See "Hyper and Deep Attention: The Generational Divide in Cognitive Modes."

9 "Perhaps then the most trenchant environmental and postcolonial criticism in relation to climate change," Clark writes pointedly, "would be one which took up the more metacritical role of examining assumptions of scale in the individualist rhetoric of liberalism that still pervades a large body of given cultural and literary criticism. An ethic attending such work would also breach current notions of decorum, redrawing the seeming boundaries of privacy whereby, say, a critic's professed views on history, religion,

colonialism or ethics are all seen to belong in the realm of 'public' controversy, seminars, papers and conferences while the resources sequestered to that person's sole use remain a supposedly 'private' matter, with a high salary and its attendant life-style still regarded, if at all, as a matter of prestige."

10 Although, as Guillory concludes, it does seem "altogether unrealistic that we should expect those without the academy to exercise the same vigilance in relation to their pleasure as we do in the context of professional reading. But it is also very unfortunate for our society that leisured reading so often falls to the level of immediate consumption, with no other end than pleasure or distraction – or rather, that kind of pleasure which is distraction" (33).

11 See Sarah Jacquette Ray, *A Field Guide to Climate Anxiety: How to Keep Your Cool on a Warming Planet*.

12 Two book-length studies offer useful ideas for rethinking scholarly reading practices in relation to modes of reading that might align more productively with the reading experiences of students: Rita Felski's *The Limits of Critique*, a postcritical project for reading motivated by a "desire to articulate a positive vision for humanistic thought in the face of growing skepticism about its value" (186), and Kathleen Fitzpatrick's *Generous Thinking: A Radical Approach to Saving the University*, specifically "Reading Together," 82–131.

13 This version of close reading has become so routine that much like "the five-paragraph essay" close reading has become "an artifact of a successful pedagogy in the schools" (Bialostosky 112). Examples of this discourse include Nancy Boyles, who argues that essentially "close reading means reading to uncover layers of meaning that lead to deep comprehension" (36); the Partnership for Assessment of Readiness for College and Careers (PARCC), who advocate reading "to reflect on the meanings of individual words and sentences"; and, in a brief review by Lorna Collier that advocates for climate education, a case for English teachers as experts at helping students examine relevant, complex, and connected stories and look for meaning and truths in and behind the words" (33).

14 Meaning is not (and never will be) "in" or "under" or "between" words. Stephen Best and Sharon Marcus describe these assumptions succinctly in their account of symptomatic reading that "often conflates three pairs of oppositions: present/absent, manifest/latent, and surface/depth" ("Surface Reading" 3–4). "Strictly speaking," Best and Marcus explain, "these are not compatible sets of terms. What is absent is simply not there; what is latent is present but invisible, unrecognized either because it is concealed or because it is undeveloped; what is deep is fully present and thus theoretically visible, but is positioned so far down, in, or back relative to a viewer, or is so completely covered by an opaque surface, that it can only be detected by an extreme degree of penetration or insight" (4). These assumptions further reinforce the asymmetrical relation between the teacher as knower and the student as recipient of knowledge. Finally, see Mary Thomas Crane, "Surface, Depth, and the Spatial Imaginary: A Cognitive Reading of *The Political Unconscious*," in Best and Marcus, for a discussion that draws on cognitive science to interpret the metaphors of surface and depth that Jameson uses to describe symptomatic reading.

15 See Rosenblatt, *Making Meaning with Texts: Selected Essays*. For recent studies of college-level reading, that build on Rosenblatt and that question current-traditional assumptions about reading, see Hornin, Gollnitz, and Haller's *What Is College Reading* (2017), which includes a survey of quantitative and qualitative studies of reading, and Patrick Sullivan, Howard Tinberg, and Sheridan Blau, eds., *Deep Reading: Teaching Reading in the Writing Classroom* (2017), which describes a theoretical approach to reading as a "meaning-making activity" in contrast to the "impoverished and reductive" understanding of reading in curriculum and state standards. Also see Sullivan, "The World Confronts Us with Uncertainty: Deep Reading and a Threshold Concept" in *(Re)Considering What We Know: Learning Thresholds in Writing, Composition, Rhetoric, and Literacy*. Eds. Linda Adler – Kassner and Elizabeth Wardle (2019).

16 Miranda B. Hickman and John D. McIntyr write in their "Preface" to *Rereading the New Criticism* that the strategy of the New Critics, "in order to establish the distinctness of the literary – critical endeavor also contributed to this widespread understanding of their criticism as dismissing what was beyond the text: they turned decisively away from approaches in the field of English that focused on literary history, read literature for philosophical insights, or appraised literature through overtly moral criteria." For New Critics such as Ransom and Brooks, "distinctions between the 'work itself' and what lay beyond it, and between criticism and other kinds of work with respect to literature, never implied that commentary that engaged extratextual matters was without value, nor that they themselves ignored such matters altogether" (26).

17 Felski's consideration of the connection between literary studies and Actor-network theory is useful. "Actor-network theory emphasizes both the necessity and the sheer difficulty of description, of attending to an empirical world that often resists or refutes our assumptions. . . . This means taking care not to conjure textual meanings out of preexisting assumptions or explanations – honoring and detailing the singular features of a text as well as the specific routes along which it travels. Actor-network theory does not exclude the political – it is deeply interested in conflicts, asymmetries, struggles – but its antipathy to reductionism means that political discourse cannot serve as a metalanguage into which everything can be translated. The task is to account for as many actors as possible, to be specific about forms of causation and connection (which are also forms of translation), instead of hitching a free ride on a preexisting theoretical vocabulary" (740) Felski concludes that if we accept the lessons of actor-network theory, we may be "less inclined to pore over a single text to draw out its hidden plenitude of aesthetic, philosophical, or sociopolitical truth – to buy into the ethical charisma of the literary critic as privileged messenger –" as Heath Love describes it. "Yet actor-network theory also pulls out the rug from under the sociologist's dispassionate analysis of a literary system: from such a bird's-eye view, everything looks remarkably similar, things blur together, and essential details are lost. Neither close reading, then, nor distant reading . . . an approach not grounded in the revelatory value of a single work or in a general notion of society or literary system but positioned on a scale between the two. Interpretation? Yes, without a doubt – but of objects and mediations as well as literary works, a practice of lateral reading across multiple texts rather than deep and intensive reading of a single text" (741).

18 This phrase is from *Shop Class as Soulcraft: An Inquiry into the Value of Work* (6). In *The World Beyond Your Head* Crawford reiterates his broader cultural critique of education and the norms that devalue forms of work. "To reclaim the real, both in the way we encounter other people and in the way we encounter things, would have implications for education," Crawford writes: "To encounter things more directly is more fundamental than doing so through representations, so maybe we needn't regard hands-on education as second-class, and those to require it as second-rate." He then reminds us that "Very few of us are scholars by nature," after all, "and it seems strange that sitting at desks, looking at books, would become the norm of universal education" (256).

19 Ghosh elaborates on how the divergence between common interest and the preoccupations of the public sphere is mirrored in the way novels have "come to be seen as narratives of identity." The broader cultural phenomenon is that social and political narratives are quests for personal authenticity and journeys of self-discovery (127) and if the climate crisis is seen "in terms of the questions it poses to the individual conscience, then sincerity and consistency will inevitably become the touchstones" for political action and judgment (133). Ghosh concludes that the "industrial, carbon-intensive economy cannot be fought by a politics of sincerity" and what we need, rather, "is to find a way out of the individualizing in which we are trapped" (134–35).

The "irony of the 'realist' novel," Ghosh insists, is that "the very gestures with which it conjures up reality are actually a concealment of the real" (23). For "it was exactly the period in which human activity was changing the earth's atmosphere that the literary imagination became radically centered on the human" (66). In England, throughout

Europe, and in North America, the fictive partition separating Nature and Culture – even as writers and movements resisted, Ghosh adds, including romanticism, pastoralism, transcendentalism – determined the human focus of literacy and literary culture. "More than a century has passed since Bill McKibbben wrote, 'We live in a post-natural world.'" Ghosh then observes, "But did 'Nature' in this sense ever exist? Or was it rather the deification of the human that gave it the illusory apartness from ourselves? Now that nonhuman agencies have dispelled that illusion, we are confronted with a new task: that of finding other ways in which to imagine the unthinkable beings and events of this era" (33).

20 Long's elaboration of the reading group offers a generative outlook on the classroom that "The interpersonal dynamics and modes of textual appropriation at work in many reading group meetings also make them occasions for engendering a particular kind of critical reflection that has a transformative potential for individuals or for the group as a whole" (117). Long elaborates these occasions in a way that is relevant to the future of close reading I am outlining here. "A socially negotiated process of cultural reflection makes these groups – when functioning well – sites for insight and innovation in the arena of identity, values, and meanings" (120).

Works cited

Alaimo, Stacy. *Exposed: Environmental Politics and Pleasures in Posthuman Times.* Minneapolis: U of Minnesota P, 2016.

Best, Stephen, Heather Love, and Sharon Marcus. "Building a Better Description." *Representations* 135.1 (Summer 2016): 1–21.

Best, Stephen, and Sharon Marcus. "Surface Reading: An Introduction." *Representations* 108.1 (2009): 1–21.

Bialostosky, Don. "Should College English Be Close Reading?" *College English* 69.2 (November 2006): 111–16.

Blau, Sheridan D. "How the Teaching of Literature in College Writing Classes Might Rescue Reading as It Never Has Before." *Deep Reading: Teaching Reading in the Writing Classroom.* Eds. Patrick Sullivan, Howard Tinberg, and Sheridan Blau. Urbana: NCTE, 2017. 265–90.

———. *The Literature Workshop: Teaching Texts and Their Readers.* Portsmouth, New Hampshire: Heinemann, 2003.

Boyles, Nancy. "Closing in on Close Reading. *Educational Leadership* 70.4 (2013): 36–41.

Buell, Lawrence. *The Future of Environmental Criticism: Environmental Crisis and Literary Imagination.* Malden, MA: Blackwell, 2005.

Clark, Timothy. "Derangements of Scale." *Telemorphosis: Theory in the Age of Climate Change*, Vol. 1. Ed. Tom Cohen. Ann Arbor, MI: Open Humanities. 148–66.

Cohen, Jeffrey Jerome. "Anarky." *Anthropocene Reading: Literary History in Geological Times.*" University Park: Pennsylvania UP, 2017. 25–42.

Collier, Lorna. "Does Climate Change Have a Place in the English Classroom?" *The Council Chronicle* (December 2017): 33.

Crane, Mary Thomas. "Surface, Depth, and the Spatial Imaginary: A Cognitive Reading of *The Political Unconscious.*" *Representations* 108.1 (2009): 76–97. (Best and Marcus eds.).

Crawford, Matthew B. *Shop Class as Soulcraft: An Inquiry into the Value of Work.* New York: Penguin, 2009.

———. *The World Beyond Your Head: On Becoming an Individual in an Age of Distraction.* New York: Farrar, 2015.

Farrier, David. *Anthropocene Poetics: Deep Time, Sacrifice Zones, and Extinction*. Minneapolis: U of Minnesota P, 2019.
Felski, Rita. "Latour and Literary Studies." *PMLA* 130.3 (2015): 737–42.
———. *The Limits of Critique*. Chicago: U of Chicago P, 2015.
Ghosh, Amitav. *The Great Derangement: Climate Change and the Unthinkable*. Chicago: U of Chicago P, 2016.
Guillory, John. "Close Reading: Prologue and Epilogue." *ADE Bulletin* 149 (2010): 8–14.
———. "The Ethical Practice of Modernity: The Example of Reading." *The Turn to Ethics*. Eds. Marjorie B. Garber, Beatrice Hanssen, and Rebecca L. Walkowitz. New York: Routledge, 2000. 29–46.
———. "How Scholars Read." *ADE Bulletin* No. 146 (Fall 2008): 8–17.
Hayles, Katherine N. *How We Think: Digital Media and Contemporary Technogenesis*. Chicago: U of Chicago P, 2012.
———. "Hyper and Deep Attention: The Generational Divide in Cognitive Modes." *Profession* (2007): 187–99.
Heise, Ursula K. *Imagining Extinction: The Cultural Meanings of Endangered Species*. Chicago: U of Chicago P, 2016.
Hickman, Miranda B., and John D. McIntrye, eds. *Rereading the New Criticism*. Columbus: The Ohio State UP, 2012.
Horning, Alice S., Deborah-Lee Gollnitz, and Cynthia R. Haller, eds. *What Is College Reading?* Fort Collins, CO: The WAC Clearing House, 2017.
Jin, Jay. "Problems of Scale in 'Close' and 'Distant" Reading." *Philological Quarterly* 96.1 (Winter 2017): 105–29.
Latour, Bruno. "How (Not) to Put an End to the End of Times?" *Facing Gaia: Eight Lectures on the New Climatic Regime*. Trans. Catherine Porter. Medford, MA: Polity, 2017. 184–219.
———. *An Inquiry into Modes of Existence: An Anthropology of the Moderns*. Trans. Catherine Porter. Cambridge, MA: Harvard UP, 2013.
———. *Reassembling the Social: An Introduction to Actor-Network-Theory*. Oxford: Oxford UP, 2005.
———. "A Textbook Case Revisited – Knowledge as a Mode of Existence." *The Handbook of Science and Technology Studies*. Eds. Edward E. Hackett, Olga Amsterdamska, Michael Lynch, and Judy Wajcman. Cambridge, MA: MIT Press, 2008. 83–112.
Long, Elizabeth. "Textual Interpretation as Collective Action." *Discourse* 14.3 (1992): 104–30.
Long, Mark C. "Introduction: Ecocriticism as a Practice of Reading." Ed. Mark C. Long. Special Issue of *Reader: Essays in Reader-Oriented Theory, Criticism, and Pedagogy* 53 (Fall 2005): 4–24.
———. "The Problem of Reading, the Practice of Writing." *The Natural History of Reading*. Ed. Leroy Searle. Centralia: Gorham, 2010.
Love, Heather. "Close but Not Deep: Literary Ethics and the Descriptive Turn." *New Literary History* 41.2 (Spring 2010): 371–91.
McIntrye, John D., and Miranda B. Hickman. "Epilogue: Toward a New Close Reading." *Rereading the New Criticism*. Eds. Miranda B. Hickman and John D. McIntrye. Columbus: The Ohio State UP, 2012. 231–38.
Menely, Tobias, and Jesse Oak Taylor, eds. *Anthropocene Reading: Literary History in Geological Times*. University Park: Pennsylvania UP, 2017.
Merola, Nicole M. "Archives of Ecocatastrophe; or, Vulnerable Reading Practices in the Anthropocene. *American Literary History* 30.4 (Winter 2018): 820–35.

Middleton, Peter. *Distant Reading: Performance, Readership, and Consumption in Contemporary Poetry*. Alabama UP, 2005.
Nixon, Rob. *Slow Violence and the Environmentalism of the Poor*. Boston, MA: Harvard UP, 2011.
Oppermann, Serpil. "The Scale of the Anthropocene: Material Ecocritical Reflections." *Mosaic: An Interdisciplinary Critical Journal* 51.3 (2018): 1–17.
Partnership for Assessment of Readiness for College and Careers. *PARCC Model Content Frameworks: English Language Arts/Literacy Grades 3–11*. 2011. Web.
Ray, Sarah Jacquette. *A Field Guide to climate Anxiety: How to Keep your Cool on a Warming Planet*. Berkeley: U of California P, 2020.
Rigby, Kate. *Dancing with Disaster: Environmental Histories, Narratives, and Ethics for Perilous Times*. U of Virginia P, 2015.
Rosenblatt, Louise M. *The Reader, The Text, The Poem: The Transactional Theory of the Literary Work*. With a New Preface and Epilogue. Carbondale: Southern Illinois UP, 1994.
———. "The Transactional Theory of Reading and Writing." *Making Meaning with Texts: Selected Essays*. Portsmouth: Heinemann, 2005. 1–37.
Sachs, Jonathan. "Slow Time." *PMLA: Publications of the Modern Language Association* 134.2 (2019): 315–31. New York: MLA.
Searle, Leroy. "New Criticism." *The Johns Hopkins Guide to Literary Theory*, 2nd edition. Eds. Michael Groden, Martin Kreiswirth, and Imre Szemal. Baltimore, MD: Johns Hopkins University Press, 2005. 691–98.
Smith, Barbara Hernstein. "What Was 'Close Reading'?" *Minnesota Review* 87 (2016): 57–75.
Sullivan, Patrick, Howard Tinberg, and Sheridan Blau, eds. *Deep Reading: Teaching Reading in the Writing Classroom*. Urbana, NCTE, 2017.
Wallace, Molly. *Risk Criticism: Precautionary Reading in an Age of Environmental Uncertainty*. University of Michigan P, 2016.

6

POSTCOLONIAL ANTHROPOCENE AND NARRATIVE ARCHEOLOGY IN *BURMA BOY*

Senayon Olaoluwa

The "Anthropocene" proliferates in the environmental humanities, even as researchers continue to engage with the relevance of its epistemic value across a broad range of disciplines. The concerns associated with the term dramatize our common humanity and the importance of erasing the false dichotomy between humans and the rest of nature. According to Christopher Wright *et al.*, "By eliminating the Cartesian categorical opposition of humans and nature, at the same time as revealing the far-reaching influence of the former, the Anthropocene demands new reflection on ontological, epistemological and methodological assumptions in Western society and the development of more integrated approaches to knowledge production" (548). But, as I will argue, it is equally, or even more, urgent that we interrogate the uncritical lumping of humans into the strain of culpability of the Anthropocene. The Atlantic Slavery that began in the Fifteenth century was the first step in the extractive logics and actualities of the Anthropocene, in which the depredation of natural resources in certain parts of the world, such as Africa and the Americas, was designed to coincide with the extraction of human resources as slaves. The tendency to exclude the colonial period from the history and rendition of the Anthropocene (Krieg 162) is grossly insensitive to the period of colonialism that conflated African humanity and nature, and subjected both to ruthless capitalist extraction, while whitewashing these horrors with the attainments of modernity.

In her book, *A Billion Black Anthropocenes or None*, Kathryn Yusoff argues that the culpability of humanity must begin with the recollection that, historically, "both slaves and gold have to be materially and epistemically made through the recognition and extraction of their inhuman properties" (3). Geology as an extractive mechanism describes the violent removal of materials for refinement and use. It is in this sense that both the mining and refinement of gold and the forced dislocation of slaves from Africa to various parts of the world served the interest of what she calls "White Geology" (4). In Yusoff's words, "To trace racial matterings across the category of the inhuman, and specifically the traffic between the *inhuman as matter* and the *inhuman as a race*, is to examine how the concept of the inhuman is a connective hinge in the twinned discourses of geology and humanism" (5, emphasis in original).

Biyi Bandele, the Nigerian diaspora writer, offers a telling insight into the nature of the Anthropocene in the novel *Burma Boy*, which, I will argue, "probe[s] the convergence of human experience and geological or climatological processes in times of climate change" (Caracciolo *et al* 221). Coming to terms with the location of his fiction in the broader context of the Anthropocene must begin with paying attention to the specific history it evokes, a critical approach that resonates with Neil McCaw's radical notion of close reading that insists on "reading to be broadened to take into account a much wider sense of culture and knowledge" (25).

The period of World War II is, for Africa and many other countries in Asia and the Caribbean, a colonial moment during which the British Empire sought to impose its wishes upon the colonized without giving them the option of resistance. In tandem with the extraction of African natural resources to serve the British interest, "human resources" were also extracted through the enlistment and conscription of colonized Africans into the Army (Gennaro 2020). The Royal West African Frontier Front and King's African Rifles alone were said to have conscripted and enlisted 500,000 men (Bandele n.p.). A huge number of the troops were lost in the war, to say nothing of those who were deformed or diseased. In this way, Atlantic Slavery and colonialism contribute to the continued deformation of Black subjectivity and the conscription of its humanity into an inhuman intimacy with the inhuman earth (Yusoff vii).

According to Elizabeth DeLoughrey *et al.*, postcolonial studies foreground "how the history of colonialism necessitates the imbrication of humans in nature" (1). At the same time, these overlapping edges do not meet neatly, do not *close*, do not suture: the shocking consequences of the ruthless extraction of natural and human resources still manifest in the subtending of human feelings to inhuman nature (Kuchta 1). I reject "the new universalism of species thinking" (Baucom 9) precisely because it was not the whole of the human species that was responsible for a phenomenon like the Industrial Revolution (Albrecht 36). New universalisms, like old ones, pinpoint the salience of Robert Young's insistence that "postcolonial remains" (19). In this close reading of Bandele's *Burma Boy*, the Anthropocene foregrounds the impossibility of its closure through a strategy I have termed "narrative archaeology."

The counterbalance of narrative archeology

The work of geology, manual and mechanical, operates not only through extraction but also through conscious and unconscious covering, and submergence of extractive remains at the site of exploitation. This is what I call *the afterlife of geology*. To come to grips with the totality of geology, as Yusoff has it, the active extraction and submergence of geological remains deserve critical attention as a cumulative reality. To this end, I bring Yusoff's notion of "White Geology" and Young's concept of "postcolonial remains" into a conversation. This confluence facilities a new and productive way of reading postcolonial literature as "narrative archeology." The counterbalancing agency of this reading consists of the archaeological resistance to the finality of burial through the persistence of excavation to discover and rediscover postcolonial remains. That is, beyond Yusoff's notion of "insurgent geology" (87), the symbolic values of excavation in archeology stand to offer new understandings of the Anthropocene from a postcolonial perspective. Narrative archeology departs from the normative understanding of the term – which specifies the material connection with the excavated artifacts in situ (Lesure 57) – by granting agency to narrative and imagination as the work of memory in fiction/art and the deployment of same as symbolically evocative of excavation tools in archeology.

Narrative archeology is, therefore, a literary/artistic inquiry to discover and rediscover the otherwise buried remains of colonial violence with its numerous resonances for the present and the future in the postcolony. The iconic work of excavation parallels the extraction of White Geology by the very sense in which it refuses acquiescence to the finality of the submergence of geological sites and their remains. It is this persistence in contesting the finality of what is buried that informs the work of archeology as a counter-disciplinary strategy to White Geology in the postcolonial Anthropocene. For archeology, time is never too distant. The evils of slavery and colonialism, which tend to be tempered by the passage of time, are often recollected in literature and arts in ways that make the distant past resonate with the present and future. Burial in the context of archeology is natural but also, simultaneously, in need of interpretation. Similarly, the political evolution from colonialism to postcolonialism and self-governance contains the remains of the colonial – as trace, text, cloud, or body in need of interpretation. There remain the symbolically protruding legs of the corpse, to echo a Yoruba adage, for which the archaeological insistence on excavation through narrative has made the question of postcolonial remains germane to the discursive age of the Anthropocene.

Bandele's writing has been read from different perspectives by scholars and critics. While some categorize his work within the mainline of "anti-neocolonialism" (Negash 77), others understand his work from an angle of migrancy and diaspora (Chris Ouma 27). His earlier fiction, especially *The Man Who Came in from the Back of Beyond* and *The Sympathetic Undertaker and Other Dreams*, seems to typify the intersection of postcolonialism and postmodernism (Ayo Kehinde 1). His earlier work has stood out in the critique of the African, and especially Nigerian, postcolonial contradictions, and this consistent inquiry, including his film productions, informs *Burma Boy*. For this reason, the novel is often read as part of the child soldier African fiction boom at the turn of the twenty-first century, which includes Ahmadou Kourouma's *Allah Is Not Obliged*, Uzodinma Iweala's *Beasts of*

No Nation, Emmanuel Dongala's *Johnny Mad Dog*, Chimamanda Ngozi Adichie's *Half of a Yellow Sun*, and Helon Habila's *Measuring Time*. While these narratives mainly unfold around African postcolonial imaginaries, Bandele reimagines the colonial past to illustrate how the contemporary abuse of children in African civil wars grows from the memory of colonialism, particularly that of the Second World War.

Disaggregating the "species thinking" of a falsely universalizing Anthropocene means, in this context, locating Britain as the "extractor" responsible for the wasted fate of Nigerian child soldiers during World War II. In Nigeria as elsewhere, "Far from being the outcome of a generic *Anthropos* . . ., the tragedy of the Anthropocene was initiated and remains driven by a powerful and unethical subset of society, whether understood as the capitalist class, oligarchs, or Anglo-Saxon nations, to name some of the groups held primarily responsible for destroying the habitability of the planet" (Wright 459). Bandele is quite conscious of such culpability when he makes King George a central force in the prosecution of the war in his narrative; King Gorge's official and authoritative personality as the head of the British Empire is incrementally repeated in the novel for emphasis. Written and pronounced in the dominant Northern Nigerian accent, King George becomes, for the most part, Kingi Joji (42, 43, 93, 160). This naming, together with the pre-eminence of the authority it commands in the Nigerian colony, allows for the performance of narrative archeology that pays attention to the depth, the reality on and in the ground, rather than the façade of the prestige associated with being recruited "to fight for Kingi Joji" (93).

By staying alive to the value of string and stakes in archaeological digs, Bandele's narrative pays attention to the primacy of grid mapping to track the contours and topographies of spaces from which colonialism has extracted minors for the Royal West African Frontier Force, children believed to be "men" fighting for Kingi Joji on the side of the Allied Forces. The narrative becomes more specific in the way it draws attention to the Nigerian Regiment as the central focus of the novel. The process of the excavation unfolds through a careful description of the spaces of geological exploitation as we read about "Saminaka in Northern Nigeria" where Farabiti Banana was born (28). The mention of a special regiment named the Nigerian Regiment is also instructive, reminding us of the particularity and peculiarity of colonial experiences, for which Stuart Hall reminds us that we are "not equally post-colonial" (242). The specificity of history and colonial experience informs why Banana is sent from GHQ West Africa in Chiringa to join 12th Battalion, Nigeria. Bandele digs further to reveal that those who have come "to fight King George's war . . . 'speak Hausa, some Ibo, some Yoruba, and some Tiv, and some speak other languages'" (70–71). Not without agency, despite the racism of White Geology, child soldiers inscribe self-affirmation in their domestication of English ranks, in which Private becomes "Farabiti" (33), Captain becomes "Kyaftin" (39), Corporal is "Kufor" (66), Sergeant Major becomes "Samanja" (37), and General becomes "Janar" (61). This renaming resonates with Said's (1983) notion of "traveling theory," whereby mobile cultures gain and lose meanings in their new environments. All these moments serve to map and redraw a distinct space of colonial exploitation and extraction of African humanity. Here, narrative archeology doesn't uncover artifacts, but the protruding and decaying legs of the colonial corpse – the enormity of colonial violence meted through White Geology hurriedly buried through political evolution. The celebration of the Royal West African Frontier Force as a colonial formation made up of able-bodied men in the service of the Empire buries the remains for Empire. As historical scholarship has shown, this formation provided the platform for the establishment of various national security formations in the preparation for political independence in West Africa (Ukpabi 485).

The glamour of fighting for Kingi Joji is displayed in the various acts of showmanship during, and shortly after, the recruitment of so-called soldiers (43). Probing into the workings and dynamics of deceptive recruitment facilitates an examination of the different layers of the colonial burial, the horrors of colonialism on the colonized. This begins with the realization that and even when fighting for Kingi Joji is ostensibly about the recruitment of "men," it is children in the main that are deceived into enlisting. For, if there is any story in Zaria that is worth listening to, it "is that Kingi Joji, monarch of Inglia, is fighting a war in a land called Boma and he wants our help" (42). At other times, the desperation of the Empire to extract the minors for the war manifests in the willful glossing over of the status of the boys as minors to make up the troops for Kingi Joji. Bandele's narrative emphasizes the overwhelming presence and enthusiasm of children; he pays particular attention this evidence of postcolonial remains, in itself and as a root of contemporary child abuse in violent conflicts in postcolonial Africa. Echoing Pulido, Krieg remarks that "life in the Anthropocene is split along an axis of 'differential vulnerability' as racialized subjects face both disproportionate geographic and material risks and expropriated benefits" (163). For the Nigerian children-turned soldiers recruited to fight for Kingi Joji in *Burma Boy*, their vulnerability is doubly so – as boys saddled with the responsibility of soldiering in Burma.

Besides being categorized as "inhuman" others by Empire, their vulnerability is concomitantly apparent in being underage, making them pliable materials for extraction to the war in Burma. Perhaps the most poignant case is that of Farabiti Ali Banana, who goes through deceptive recruitment and training in Kaduna; he is just a little over 12 years old and finds himself on the battlefield at 13 (47). He represents countless other underage soldiers that populate the war – in the novel, from Yusufu and Iddirisi (50) to Pash, Guntu, and Bloken (67, 158, 184).

The manipulation of underage children for recruitment takes an extreme dimension in Gboko. Conscription plays out like an extended Christmas party gifts to children by Kingi Joji, considering how Tiv boys are invited by the District Officer for "the king's shilling" in what seems an expression of solidarity with the indigenous celebration of the New Yam Festival in Gboko (184). Taking advantage of the reputation of the Tiv as a "martial race," little children like Bloken are deceived to become fighters for Kingi Joji. Yet all this is played down by Damisa's response to Banana's remorsefulness about having lied regarding his age to be recruited. By declaring that "[w]anting to be a man is no sin" (47), the Sergeant Major becomes complicit in the extraction of children as material for wastage, as literal cannon fodder, under the pretext of fighting for their (English) king. This becomes clear at the onset of actual battle in the march toward Mawlu; Damisa admits that, apart from Zololo, the entire troop, the D-Section in the column he commands, "were all children" [and . . .] it wouldn't hurt them to be reminded" (110). If the rendition of the Royal West African Frontier Force in history has so far served to bury the remains of the atrocities regarding the abuse of African children as soldiers, Bandele's narrative archeology – itself a writing that close reads history – reveals the insufficiency of such covering.

According to Caracciolo *et al*, Anthropocene fiction has come to be associated with a particular deployment of metaphors; they explain that "metaphor plays a key role in closing the gap between everyday, embodied experience and more intangible or abstract realities – including, we suggest, the more-than-human temporal and spatial scales that come to the fore with the Anthropocene" (221). Bandele's fiction is keenly alive to its own use of imagery and metaphor to embody the experience, the foreboding, of child soldiers. At the beginning of the trudge through the Burmese jungle, Banana realizes how trapped they all are, because as only he can see, a "swirling inferno appeared in the distance, a blue pillar of fire roaring towards them" (25). Having been previously calmed by the Janar (General) in the face of even the fiercest of hostilities from Japanese soldiers, this seems like a suicide mission, and it is instructive that the Janar's speech ends with "Good luck and Godspeed" (27). It provokes Banana's critical reflection on whether "walking quietly and confidently into the angry flames would earn them the Janar's esteem if they met him in the hereafter" (27). The metaphor of the "inferno/flames" speaks not only to the danger faced by child soldiers, but to the nature of their destruction: like gold, mined and re- or unmade through flames. This imagery resonates with an idea of Anthropocene as world-maker/destroyer of worlds (Yusoff 16). The fate of Banana, in particular, is symbolized by recurring encounters with flames in the narrative, such as this one, in which a great many small flames combine into one destructive force: "a million fireflies fused together into one giant piercing flame" (81). Despite the beauty of such an image, it is horrific, a reminder that this "piercing" spares neither humans nor nature when the violence of the war comes full circle.

In *Burma Boy*, petrol discourse is constitutive of the confrontation with the dominant narrative of World War II through processes of the mobilization, operation, evacuation, dispersal, and demobilization of troops. The agency of fossil fuel is the driving force, evinced by the novel's cover illustration in which, beside a soldier holding a riffle, two fighter jets are suspended in the horizon as the "empire of machine" at the center of the war. Bandele seeks out a moment illustrative of the prelude to the warfare to contrast with the destructive capacity of Empire:

> The going was not easy at first. They sauntered through a village and startled a group of girls washing by a stream. The girls smiled shyly and

> then tore off into the rice fields as soon as the men smiled back. . . . They stopped at the stream to fill their chaguls. . . . A herd of buffaloes was guzzling away downstream. 'If it's good for them I reckon it's good enough for us,' Samanja Show declared, which seemed perfectly sound logic.
>
> *(89)*

The picture painted here is that of harmony of the human and the natural at the threshold of imminent violence. Significantly, the soldiers are only able to help themselves with drinking water from the river in the absence of a Medical Officer (MO) because the buffaloes have stepped in to assure them that the water is safe. This gesture of assurance in itself dramatizes the imbrication of human and natural lives. But the bevy of shy girls who playfully makes for the rice fields together with the appearance of the column of soldiers presents a scenario of both freedom and bondage. Whereas the girls are at liberty to depart the river at will, the young soldiers – children in the main – are restrained by the regimen of the warfare, proverbial lambs led to the slaughter.

Using the topography of the excavation, Bandele draws attention to the double vulnerability of the Nigerian minors deceptively conscripted into the service of Kingi Joji. The Allied Forces were not an impregnable formation, and their vulnerability is manifested in various ways, including the crash that killed the Janar (75) alongside other crew members. But more important is the toll war takes on members of the Nigerian D-Section. Though Pash is injured and withdrawn for amputation in the early part of their march, his fate seems mild compared to those of the other members. After weeks of the nightly killing of Japanese soldiers, the overwhelming putrefaction of bodies and the pollution of the river lead to a malaise of pathologies. The march to Nyaunggaing, estimated to be in an hour, turns out to be terminal for most members of the D-Section. Guntu and Ko, for instance, are felled by the "trip wire of a mine" when "seconds later several luminous metallic fruits dropped out of the trees above . . . and exploded into a dense constellation of rattling bullets" (192).

Returning to Banana, his brutalization is reinforced by witnessing the vulnerability and fate of the D-Section. In quick succession, he watches Danja and Gillasfsie "reel and topple over as they were hit" (192). In the process of saving Banana during the crossfire, Damisa is hit and badly injured. Although Damisa saves Banana's life, Banana must obey the order to shoot Damisa dead, and is subsequently saddled with the responsibility of lifting his lifeless body and easing him "unto the swollen waters" (203). Although shooting Damisa is an act of euthanasia, a way to avoid the Japanese, who are notorious for dismembering wounded members of the Allied Forces (200), Banana carries out the order while wiping "tears off his cheeks" (202). Much as *Burma Boy* belongs to a category of African novels that "attempt to reflect the use and abuse of minors as active agents in warzones" (Gray 4), it is also about the consequences of this abuse, the immediate and ongoing traumatization of child soldiers and black postcolonial subjectivity. As the novel winds toward the conclusion, Banana is seen returning stark naked to

White City after he has donated both Browning and Enfield revolvers to animals with whom he has had a monologue, having lost his mind (207).

Banana's insanity on the heels of killing Damisa provokes a "decision" by the natural elements who maintain "a healthier distance" from the scene:

> A flock of brown dippers sunbathing on a rock nearby screeched in alarm and dived to the bottom of the water where they stayed for a whole minute. When they emerged from the belly of the stream they had established a healthier distance between themselves and the frightening sound. A curious wild dog that had been watching the two men all morning growled angrily and scampered off.
>
> *(202)*

The angry growling of the wild dog punctuates the limits of human violence, suggesting that the question of environmental justice is farther than the reach of human agency (Olaoluwa 2020). The "healthier distance" maintained by nature takes a toll on Banana; this disavowal of nature may be read as a symbolic depiction of the "culpability and guilt of child soldiers in fiction" (Sanders 195). Banana's striking, repetitive dreams about the exchange of garbs between Kingi Joji and the Emir of Zaria (160, 205) suggest child soldiers' anger and disappointment as well as their capacity to see through the collusion of colonial oligarchs. The epiphany serves to announce the transition of the children from innocence to knowledge at a stage that is rather too late, having been told in their dreams by the oligarchs that the war is over only for them to wake to no such news. It leaves them to wonder about the fate of the world "when kings and emirs can no longer be trusted to tell the truth" (160). *Burma Boy* is more than another African fiction of child soldiers at the turn of the twenty-first century; Bandele's narrative digs have uncovered the complex layers of the foundations of the postcolonial challenges on the continent.

Using the conceptual framework of narrative archeology, this chapter suggests the impossibility of closure to the memory of World War II in the Nigerian postcolony, literalizing the Yoruba metaphor of the protruding legs of a hurriedly buried corpse. By shooting out from the grave, not only does protrusion invite the curiosity of the archaeologist, it also symbolically signals a projection to the future, underscoring the perpetuation of colonial violence in the Anthropocene. The violation of young children and their conscription as soldiers in postcolonial Africa during civil wars hark back to the initial conscription of children during the colonial era of World War II. Close reading Bandele's narrative archeology serves not only to excavate this improper colonial burial of the past but also the present humanist fiction buried in a universalizing notion of the Anthropocene.

Works cited

Adichie, Chimamanda Ngozi. *Half of a Yellow Sun*. Knopf, 2006.

Albrecht, Monika. *Postcolonialism Cross-examined: Multidirectional Perspectives on Imperial and Colonial Pasts and the Neocolonial Present*. Routledge, 1991.

Bandele, Biyi. *Burma Boy*. Farafina, 2007.
Bandele-Thomas, Biyi. *The Man Who Came in from the Back of Beyond*. Heinemann, 1991.
———. *The Sympathetic Undertaker, and Other Dreams*. Heinemann, 1993.
Baucom, Ian. "History 4°: Postcolonial Method and Anthropocene Time." *Cambridge Journal of Postcolonial Literary Inquiry*, vol. 1, no. 1, 2018, 123–42.
Caracciolo, Marco, *et al.* "Metaphorical Patterns in Anthropocene Fiction." *Language and Literature*, vol. 28, no. 3, 2019, 221–40.
DeLoughrey, Elizabeth, *et al.*, editors. *Global Ecologies and the Environmental Humanities: Postcolonial Approaches*. Routledge, 2015.
Dongala, Emmanuel. *Johnny Mad Dog*. Translated by Maria Louise Ascher. Picador, 2006.
Gennaro, Michael. "'The Cause Is a Worthy One, So Come Along with Your Sixpence and Enjoy Yourselves with One Hour of Lusty Sport': Sport in Lagos, Nigeria during WWII". *Journal of African Military History*, vol. 4, nos. 1–2, 2020, 41–65.
Habila, Helon. *Measuring Time*. Hamish Hamilton, 2007.
Hall, Stuart. "When Was 'the Post-Colonial'? Thinking at the Limit." Iain Chambers and Lidia Curti, editors. *The Postcolonial Question: Common Skies, Divided Horizons*. Routledge, 1996, 242–59.
Kehinde, Ayo. "A Parable of the African Condition: The Interface of Postmodernism and Postcolonialism in Biyi Bandele-Thomas's Fiction." *Revista Estudios Ingleses*, vol. 16, 2003, 1–36.
Kourouma, Ahmadou. *Allah Is Not Obliged*. Translated by Frank Wynne. Heinemann, 2006.
Krieg, Parker. "Planetary Delta: Anthropocene Lives in the Blues Memoir." *Auto/Biography Studies*, vol. 35, no. 1, 2020, 161–81.
Kuchta, Todd. "Reading *Reef* in the Anthropocene." *South Asian Review*, 2020. DOI: 10.1080/02759527.2020.1829442. Accessed 20 November 2020.
Lesure, Richard. "Emplotment as Epic in Archaeological Writing: The Site Monograph as Narrative." *Norwegian Archaeological Review*, vol. 48, no. 2, 2015, 57–74.
McCaw, Neil. "Close Reading, Writing and Culture." *New Writing*, vol. 8, no. 1, 2011, 25–34.
Negash, Girma. "Migrant Literature and Political Commitment: Puzzles and Parables in the Novels of Biyi Bandele-Thomas." *Journal of African Cultural Studies*, vol. 12, no. 1, 1999, 77–92.
Olaoluwa, Senayon. "Dislocating Anthropocene: The City and Oil in Helon Habila's *Oil on Water*." *ISLE: Interdisciplinary Studies in Literature and Environment*, vol. 27, no. 2, 2020, 243–67.
Ouma, Christopher. *Childhood in Contemporary Diasporic African Literature*. Palgrave, 2020.
Said, Edward. *The World, the Text and the Critic*. Cambridge, MA: Harvard University, 1983.
Sanders, Mark. "Culpability and Guilt: Child Soldiers in Fiction and Memoir." *Law & Literature*, vol. 23, no. 2, 2011, 195–223.
Ukpabi, Samson C. "The Origins of the West African Frontier Force." *Journal of the Historical Society of Nigeria*, vol. 3, no. 3, 1966, 485–501.
Uzodinma, Iweala. *Beasts of No Nation*. John Murray, 2005.
Wright, Christopher, *et al.* "Organizing in the Anthropocene." *Organization*, vol. 25, no. 4, 2018, 455–71.
Young, Richard. "Postcolonial Remains." *New Literary History*, vol. 43, 2012, 19–42.
Yusoff, Kathryn. *A Billion Black Anthropocenes or None*. University of Minnesota Press, 2018.
———. "Anthropogenesis: Origins and Endings in the Anthropocene." *Theory, Culture & Society*, vol. 33, no. 2, 2016, 3–28.

7

KEY WEST IN THE ANTHROPOCENE

Stevens and Bishop close reading Florida

Peter Balaam

The traditions of American poetry well up to the twentieth century assumed a relation between human mind and the material universe that was typical of the late Holocene era. In this relation, human observers took the world's material web of forms and phenomena as a language, a mirror, a goad for thinking and imagining, a shore to shove off from. Snowstorm, sunrise, bird, tree, and mountain – all drew the eye and other senses, and the mind drew the world *back* again. The impulse was much older than America. As Adam's first task in Eden was naming the animals, so, presumably, it has been assumed to be the task of *Anthropos* to take the wide world as there to be *read*, named, and known. Natural objects and phenomena so perceived were at once more than merely themselves. Whatever else they might be on their own, they served as well within a system of relations and bore ideas and correspondences to human thinking, feeling, and meaning-making. But what happens to a poetic tradition with roots in the previous geological era as we find ourselves now within the Anthropocene, this era luridly named for ourselves, and in which we find human imagination burdened with the knowledge that the actions of a single species, our own, has become the dominant geomorphic force within the Earth system?[1] How may the material universe in which our senses evolved remain a resource for the imagination when, as Dale Jamieson and Bonnie Nadzam inquire, "nature" has become a field thoroughly subjected to human management (11–13)? What happens to human knowing when, as Clive Hamilton and Bruno Latour teach us to fear, disrupted forces within the earth system stir and waken, no longer serving as material backdrop to human aims, assumptions, and preoccupations?

It was to notions of *correspondence*, the idea of mind in linguistic relation to nature, that, in a 1995 article, "Wallace Stevens and Elizabeth Bishop: Ideas of Order," Albert Gelpi turned to compare the twentieth-century Key West poetry of Wallace Stevens (1879–1955) and Elizabeth Bishop (1911–1979). Stevens's 18 years of annual winter visits to the island, and Bishop's 11-year residence there overlapped in the 1930s, making Gelpi wonder what, had they encountered each other on the beach, the two poets would have had to say to each other (157). Though a full generation apart, he found, the poets held much in common. Both exhibited Modernist skepticism and an "existential sense of living in a world of chance," and both balanced such views with more continuity with the Romanticist past than is assumed. They shared an admiration for Samuel Taylor Coleridge and the lifelong habit of working out Romantic conceptions of the imagination for chastened twentieth-century sensibilities (164).[2] Most important, for Gelpi their Key West poems revealed how "the pragmatist and idealist aspects of their sensibilities" and by extension that of broader American poetic tradition seemed to meet and jostle along the beach, that paradigmatic line for both poets where land meets water. Comparing their Key West poetry, Gelpi discerned in the differences as one moved from Stevens to Bishop evidence of the "decline" from modernism to post-modernism; a revelation, he observed, that made *their* relation more important than the one more often noted between Bishop and her mentor, Marianne Moore. The decline Gelpi describes descends from Stevens's "heroic aspiration that the imagination can help us prevail to [Bishop's] wry hope that it can help us at least to endure and, in the poets since Bishop and Lowell, to the bemused acceptance that it can only help us to play" (164). But how do such views matter as we find ourselves in the Anthropocene, where, with enormously expanded stakes, we contemplate not just our own evolved relation to external nature but respond to its radical diminishments, both present and coming? What have older and earlier poets to teach us about the encounter of mind and nature in this era that *Anthropos* has made in its own image? The Florida poems of Stevens and Bishop offer a compelling portrait of modern Anthropos confronting the scene of diminished nature and close reading it. They serve as able guides to several disciplines—of attention, descent, and losing—that ought to typify that encounter.

In a recent anthology, *Can Poetry Save the Earth?* John Felstiner asks "Realistically, what can poetry say, much less do about global warming, seas rising, species endangered?" His equally realistic reply, "Next to nothing," sets up a defiantly mathematical exception, however small: "Yet next to nothing would still be something" (7). To think realistically – that is, modestly – about the value of close reading in the Anthropocene might begin in asking whether the stance of open attention lying at the heart of close reading might resonate with paradigms of ecological thinking. As generally practiced, close reading was ever less about reading for unity, ambiguity, and irony than it was a rationale for and an invitation to heed literary art's invitation simply to notice. According to Cleanth Brooks's classic statement in "The Language of Paradox" (1947), John Donne's intricate conflation of divine and carnal love in "The Canonization" produces an image of erotic striving on par with and parodying canons of Christian sainthood such as readers habituated to the instrumental language of modern scientific discourse might barely register. And how can such readers value what they cannot perceive? When we speak of "close reading" as a practice, we misperceive if we imagine the kinds of labor inherent in it – scrutinizing attention to the text, recognizing patterns, disregarding the extrinsic – are the work only of the reader. Such concerns are the work of the poet as well, and also of the poem. Brooks's larger point was that the poem, itself, in the demands of its particularity, teaches readers how to read, how to notice, how to value. As Andrew Dubois more recently observes, analysis reveals that not just the reader but "the literary object itself is already doing critical work, that reading is not just what one does to a poem, but that it is often already happening inside the poem" (16).

Indeed, though the New Criticism held no truck with environmentalist practice, its invitation to take attention itself seriously aligns with William Cronon's observation that to preserve nature, we must start with the one in our heads (22; qtd. by Felstiner 6). Jane Gallop observes that in an era that had grown suspicious of grand narratives, close reading was "tarred with the elitist brush applied in rejection of the New Critics' canon" with its ostensibly "timeless universals." But, Gallop asserts, hailing close reading as an irreplaceable way to teach the practice of paying attention, "it is precisely my opposition to timeless universals that makes me value close reading" (185). As the Anthropocene sets us before the rise of *material* "universals" in insistent global form, close reading could yet contribute much to necessary acknowledgment of what is happening, not only guiding our reception of changing conditions on the planet, but also, offering clearer reckonings of our roles in bringing about such changes and of our perceived and actual place in a world so altered.[3]

Today's Florida Keys are the remnants of ancient limestone formations and coral reefs formed under the sea during the Pleistocene and exposed to the air only 100,000 years ago as ocean levels dropped during the last Ice Age. The string of nearly 900 charted islands swings to the south and west from the southeastern tip of the Florida peninsula into the Gulf of Mexico. Key West is the southernmost of the group and the furthest west with adequate freshwater to sustain human settlement of any size. The name derives from the englishing of the Spanish for "bone island" – *Cayo Hueso* – that name itself a likely misnomer given in reference to the piles of bleached-white and bonelike staghorn corals that litter the beaches throughout the group. Acquired by the United States from Spain in 1822, Key West went through cycles of economic boom and bust in the nineteenth century and had an attractively sleepy and forgotten air when in the 1920s it became the Prohibition-era playground of Ernest Hemingway and John Dos Passos, defectors from their roles in "Lost Generation" Paris. That Key West – off-grid and lying far beyond effective federal control over traffic in rum – offered superb fishing, cheap housing, and a pleasingly "not quite American" scene to the writers and other artists who soon followed and contributed to Key West's twentieth-century reputation as an art colony.

Stevens made his first visit to southeastern Florida in 1916 in connection with his role as a lawyer for the Hartford Accident and Indemnity Company, and soon established an annual winter visit of several weeks to Key West that he kept unbroken from 1922 to 1940. Bishop, together with her partner, Louise Crane, discovered Key West in 1938, buying a house there, and making it her home for 11 years. Both poets approached this remote island along the Gulf Stream with a sense of transport, delighting in its physical beauty and comfort, and availing themselves of the excellent swimming, fishing, and relaxed social mores it offered. While Stevens's letters home from his earliest trips to Florida reveal his enthusiastic response to the subtropical scene of beauty, as is well-known, with the 1936 poem "Farewell to Florida" he shook the Gulf coast from his poetic sandals for good. But from the start Stevens's earliest poems register Florida as a "seductive" but ultimately "ambiguous earthly paradise" (Scroggins 38). The brief "Indian River" from 1917, describes the Atlantic-coast "inland waterway" as an intricately synesthetic scene so eye popping it sets a bright musical note vibrating:

> The trade-wind jingles the rings in the nets around the racks
> by the docks on Indian River.
> It is the same jingle of the water among the roots under the
> banks of the palmettoes,
> It is the same jingle of the red-bird breasting the orange-trees
> out of the cedars.

The strangeness of visual details initially seems simply wondrous, especially in these long lines of insistent anapests. The rhythm is the effect of strings of prepositional phrases that locate the source of the jingle "in the nets around the racks by the docks on" the river. The next lines oddly produce "the same jingle" as the effect of the "water" intricately located "among the roots under the banks of the palmettoes." Despite the trouble gone to, however, the sentences gain no ground: "It is the same, . . . it is the same." Stevens's Florida is not merely prescient of the Anthropocene, but already expressive of it: every aspect of the landscape along Indian River is already "human" in construction, import, or intention. A "jingle," innocuous vibration, is also a marketing ditty. The wind, Romantic image of spirit *par excellence*, is at Indian River the "*trade*-wind" and the musical string of plural monosyllables "rings . . . nets . . . racks . . . docks" produces the tackle of fishing and commerce piled to dry on a wharf. A landscape with *orange* trees seems exotic or perhaps surrealist, but these are "orange-trees," celebrated engines of Florida's commercial economy. Even the adjective "Indian" in both the place name and the poem's title flatly marks and replaces any particular indigenous ways that this landscape was once differently peopled. The intricate visuality of the line "the red-bird breasting the orange-trees/out of the cedars" seems to shove off somehow from Coleridge's account in "Frost at Midnight" of spring as the time of year when "the redbreast sit and sing/Betwixt the tufts of snow on the bare branch/of mossy apple-tree." That comparison helps us to see that Stevens's point is the static un-deciduousness of the scene. Instead of the momentous and regenerative change that comes to the temperate nature of northern lands, the scene at Indian River produces only and ever the same subtropical "jingle": "there is no spring in Florida, neither in boskage perdu, nor/on the nunnery beaches." The un-deciduous sameness of Florida in palmetto swamp and along the beaches he describes as "nunnery" (somehow locked up, unfruitful) renders the scene at "Indian River" a jaunty dystopia, a stimulating but ultimately sterile sameness.[4]

In less visual and more erotic register, the speaker of "O Florida, Venereal Soil," navigates the demands of his own enticed responsiveness to the warm breezy nights of Key West. "Florida" here, personified as a goddess "Stooping in indigo gown" in the night over the sea, allures and overwhelms the poetic imagination. The speaker pleads with the goddess to narrow the sensuous panoply and "dreadful sundry" of this world and to "disclose" instead just a "few things for themselves." In these poems we see a version of the signature drive of all Stevens's work, a drive to cut through the accreted "layers of varnish" with which human culture and thinking constantly paints the world about and to encounter nature in a direct, limited, but more precise way (Hofmann). In the wonderfully titled, "Nomad Exquisite" (1923), the questing seeker after greener pastures who speaks paints the subtropical scene as a fabulous garden watered by an irrigating profusion he calls the "immense dew of Florida." The idea of moisture *falling*, condensing out of the atmosphere as dew, seems somewhat perverse in light of the Florida peninsula's unusually profuse aquifers and historical associations to helmeted conquistadors sweating in search of youth-restoring fountains. The point is the "immense[ity]" of the wet, and all that it "brings forth." Stevens compares the prodigious creative flow of Florida's subterranean aquifers to the upwelling, brimming, subterranean flow of the poetic imagination, which as fabulously draws out new "leaves" of form: *As* the dew "Brings forth/the big-finned palm,/And green vine angering for life," and as it inspires the traveler-viewer's response to the scene in "hymn and hymn," *as well as* "blessed mornings/Meet for the eye of the young alligator," just so, "in me, come flinging" products of the imagination, "Forms, flames, and the flakes of flames." The parallel riffs again on Coleridge, whose speaker in "Kubla Khan" imagines the subterranean "sacred river" of the imagination violently "flung" upward into sight. Just as having a "mind of winter" allows one to see a northern, scene with trees "shagged with ice" without indulging the temptation to identify such austerities of winter climates with the human concept of "misery," so here, acts of attention in Stevens's Florida poems consistently lead us to distinguish between the natural scene that is there and that scene's effects upon the mind, a distinguishing that opens a window upon our own cognitive and affective functioning (Stevens 47, 95, 9).

The broad, plain light of Key West delighted Bishop as well. And like Stevens in his ambivalence, Bishop's embrace of the "state with the prettiest name" had no need to idealize or to overlook its commercial and constructed valences. Living there brought about in her work a new style characterized by a sustained attention and description of the objective world about that distinguishes the earlier (or *northern*) poems in *North & South* (1946), her first collection, with their surreal scenarios of "claustrophobic introspection," from the later (southern), Key West-inspired poems in that same volume (Travisano 149).[5] The poems Bishop wrote in Key West turn to the natural scene to close read it, her narrating eye making its way around a sight or scene as if performing an ekphrastic description. Noted for their precision, for sustained acts of attention that result in a sense of emotional grounding, these poems do not merely describe but "display the mind thinking . . . slowly building to an unexpected climax" (Travisano 168).

Just such a sustained and attentive reading of the brackish swamps near Naples in the months before her first visit to Key West, Bishop's "Florida" tracks the push and pull of her earliest responses to the Gulf Coast world typical of the "state with the prettiest name." Moving quickly beyond the pretty, however, her eye registers of the place a grubby provisionality and persistent mixture. At sight, merely "float[ing] in brackish water," the scene seems "held together by mangrove roots." Her eye swings in close to what in "The Bight" 11 years later she would embrace as "all the untidy activity," both "Awful" and "cheerful." Neutral, inclusive, unafraid, her eye, lighting on the given, the found, produces gradually from them a sense of the real. Mobile, contemplative, this eye on nature resembles Darwin's on nature's "entangled bank" (211) or Ishmael's in the Arsacides, anchoring in the paradox that "Death trellised Life" (Melville 331). The mangrove roots holding the place together "bear while living oysters in clusters/and when dead strew white swamps with skeletons." If so, these bones-bestrewn swamps are dotted with green "hummocks" of regenerative life, themselves like "ancient cannon-balls sprouting grass." Other gestures of self-conscious blending and lowering follow. The bodies, calls, plumage, and behaviors of birds inspire analogies – "in a tantrum," "embarrassed," and "clown[ing]" – correspondence-thinking indulged in knowingly.[6] Though the "coast-line" sags "monotonous, endless," up close, the beaches reveal the empty skulls of "Enormous turtles, helpless and mild," their eye-sockets "twice the size of a man's." Still closer, the sands are "delicately ornamented" with "tide-looped strings of fading shells," seeds, and animal bodies that present their own intricate forms to be mixed with human names and lore, invitations to recognize, read, and master: Job's Tear, Chinese Alphabet, Junonia, Ladies' Ears, and the "Indian Princess" whose figure some see in the mottled carapace of certain crabs.

With Gelpi, we may initially conclude that Stevens and Bishop seem to pursue different projects on the beach. Stevens wants to scrape away the varnishes of human thinking that falsely deepen and sharpen our projective conceptions and so prevent what could be a rawer experience of sensing the natural world about. In "Florida," at least, Bishop may seem, on the contrary, grounded and sustained by bringing the human knowledge project of reading and naming an encyclopedic array of objects around her. But both poets step *down*, and *in*, into a scene far from the easy Anthropocentric Florida of fanning breezes and fountains of youth. The immanence and attentive inclusiveness of the speaker's gaze becomes even broader of embrace in the second stanza, which moves from the beach to the brackish wetlands beyond where buzzards drift downward like "stirred up flakes of sediment/sinking." "Woods fires" fill the air with smoke that the eye registers as "blue solvents." Whatever might be dissolving in the luminous chemistry aloft, in the swamp below all is allowed and presented. The burned stumps of vanished trees are of a "velvet black," and hordes of mosquitoes hunt to the "ferocious obbligatos" (that is, from a musical score, *requisite* accompaniment) of their own buzzing. As attuned to the necessity of mosquito *music* as she is to the swamp's givenness, variety, and reality, Bishop's ekphrastic reading of the scene breaks us to the limits that truer ways of seeing require. As a big moon rises, Florida sends invitation to a scene far from the tourist-office promotional paradise: "Cold white, not bright, the moonlight is coarse-meshed,/ and the careless, corrupt state is all black specks/too far apart, and ugly whites; the poorest/post-card of itself." The poem's studied embrace of the brackish swamps of the southwest Gulf coast subtly becomes for the poet entrée to a newly glimpsed and un-transcendent ground of poetic power. As Stevens found a metaphor for poetic power in the figure of the irrigating "dew" that could conjure "blessed mornings/Meet to the eye of the young alligator," so Bishop embraces the "untidy" scene in Florida as a step toward a more grounded, less anxiously subjective style, integrating its parts and claiming no personal stake or identification in them. Bishop's demanding ending to "Florida" gestures, perhaps, toward the young alligator's "eye" in Stevens, where that eye is analogous to the poet's and the poet quick to announce the mornings "meet." Bishop complicates Stevens by moving beyond just the visual; her ending happens "After dark" and thus appropriately the shift is from the visual to the expressive. Discriminating "five distinct calls" in the alligator's repertoire, Bishop lays no claim as a poet to some notion of reptilian affective range. Instead, more simply, and with greater restraint than Stevens, she integrates the *expressions* of the swamp (the alligator's calls) with the form of objects on the beach (the Indian Princess). Where Stevens had playfully decreed all to be meet to the eye, in Bishop all is tender and uncertain, actually, incipient; the alligator with its "five distinct calls" from the swamp, "whimpers and speaks in the throat/of the Indian princess."

The new objectivity Bishop developed in Key West drew existential grounding from the wondrous and marvelous that exist within the mundane and quotidian, as Bishop called it in a letter, "the always-more-successful surrealism of everyday life" (*Elizabeth Bishop*, 861). Such this-worldly surrealism resonates with the anthropological project of Stevens's work that, in several Key West-based poems, scrutinizes the yearnings for meaning-making started in the human mind by light, sea, and sound. With a keen sense of mischief, Stevens chastises *Anthropos* for the delusions that follow from instinctive belief in his own romantic perceptions. His perverse-seeming genre designation of *fabliau* – a bawdy tale of the French Middle Ages – for the brief *romance* in "Fabliau of Florida" is telling. What exactly is ribald about a luminescent gleam, a "Barque of Phosphor" departing "the palmy beach" on intractable adventures across a sea striped in "alabasters/And night blues?" The poem is slight, a daydream in 12 lines about movement, visual contrast, and absorption. What's bawdy about it, put-on and excessive, what makes it a *fabliau*, is the garish degree of evident artifice the mind seems willing, even eager, to entertain. It results in Stevens's deadpan observation that "There will never be an end/To this droning of the surf." A similar note of ancient ritualism surfaces in "Infanta Marina," in which Stevens personifies the spirit of the sea at twilight with a formal title from the Old World, "infanta," a daughter of the Spanish King, ineligible for the crown. The word – etymologically *in* + *fanta* – means "not-speaking," as in English "infants" do not. But, if inarticulate, the evening sea at Key West is for that no less histrionic. As the beauty-struck correspondence-thinker Emerson inquired of a glorious January sunset: "What was it that nature would say?" (*Nature* 15), so here, the twilit Infanta appears on the beach together with the Evening, her consort, and for a brief time, roams the sand in a dumb-show of plumes and grandiose gestures. Perhaps seductively but in any case ridiculously, only briefly "Partaking" of the very sea and the evening that they are, these august and costumed personages preoccupy the observing speaker, organizing the space and scene of the beach as they "flowed around/And uttered their subsiding sound." Like the endless roar of the surf, that draining noise you hear as this half-preposterous negotiation transpires is the sound of a hoped-for transcendence failing.

Stevens's Florida poems have their own verbal verdure, but they playfully expose the effect on the mind of this encounter with a lush and exotic scene. Rightly famous for its difficulty, his work is actually neither abstract nor arcane. Its project is to draw close to the real, the ordinary, what actually happens as mind confronts world, and to read *it*.[7] In this sense, the great contribution in the Anthropocene of his most famous Florida poem, "The Idea of Order at Key West," lies in its precise inquiry into the complex process of what happens as knowing and expressing mind confronts the scene of nature. Exactly what relation obtains between the poetic singer's song of the ocean she walks by and the plunging ocean there? And what is the effect of that relation upon the perceptions of others who overhear the song? It is once the singer's song has ended and the speaker and his companion turn back toward the town that things become most interesting. The previous five stanzas inconclusively frame the questions Gelpi's inquiry raised – "whose spirit" resides in the voice of the singer (as poet or artist)? What role has the singer's "song" (the expression of art) in engendering meaning, in persuading the listeners that what they hear is the "tragic gestured" sea? What role has the naturalistic reality of the sea in that meaning? What is the status of human perceptions of correspondence between mind and nature, whether they be real or illusory? The noted sixth stanza contains the speaker's direct address to his companion: "Ramon Fernandez, tell me, if you know,/ Why, . . ./. . . tell why . . ./The lights . . . // *Mastered* the night and *portioned out* the sea/*Fixing* . . ./*Arranging, deepening, enchanting* night" (my emphases). The syntax forms a clearly rhetorical question, without question mark, the asking of which makes his point: when we turned toward town, why should the lights there have had the power to "master the night" and to "portion out" the sea? In how far do our own sensory-cognitive powers effect (and not just affect) the night that we experienced? And, as if we could say, what if they didn't do that? What nature might actually be underneath the flow of that constant sensory report and mental framing can only be vaguely sensed in "ghostlier" but also "keener" sounds. Instead of an answer, what the poem delivers is the provocation of an invitation to attend to the processes of mind, to a self-consciousness about how we perceive. In the Anthropocene, such attention could be the beginning of ethics. What his unanswerable questions frame is the drive in *Anthropos* for mastery as well as that drive's invariable failure, our wishing to not be *in the dark*, our willingness to fill the sound of the wind with knowledge.

It has long been observed of Bishop that attentive and precise description of landscape in her work has the function of an existential anchor, an idea as apt for the speakers of individual poems as for the famously rootless poet, herself. Several of Bishop's speakers attend to and describe the world about in close readings analogous to that performed by the 6-year-old Elizabeth, protagonist of "In the Waiting Room" (1976), who speaks her name and age aloud and desperately fixes her eyes to the images within the familiar goldenrod-colored frame of the *National Geographic* magazine in order "to stop/the sensation of falling off/the round, turning world/into cold, blue-black space." The "obsessed" attention of Bishop's "Sandpiper" (1965), "looking for something, something, something," is a close reader in this sense as well, as is the rescued and surviving Crusoe, who asserts suddenly remembered details of his survival as *non sequiturs* with a force and specificity that seem intended to serve as a foothold in a blur of painful losses: "– And Friday, my dear Friday, died of measles/seventeen years ago come March" (*CP* 160, 131, 166). But if Bishop's relation to the natural world is about desperately holding on, it is a *holding on* with eyes and other senses open and receptive, following the lead of realities in an external world that is other to the viewer and really there. Nature in Bishop is always more than her speakers' illusions or projections. Whereas Stevens plays mischievously with perception's house-of-mirrors, Bishop borrows the material *thereness* of objects in the world without, enlisting them as anchors – real, weighted, and with *bite* enough to swing to in a universe of chance. In "Seascape," she describes an immanent watery scene of birds in flight, reflective water, mangrove islands, to charge the conceit that the scene, somewhere in Key West, "does look like heaven." Her speaker's rapt visual focus and sense of transport at the beauty of this watery world is set up against the inward-looking idealism of a clerical lighthouse, villain of the piece, who rejects altogether the speaker's empirically grounded but playful receptiveness of the scene. The lighthouse believes only in the "strong glare" that he articulates from within. Unwilling to actually look at the world about, he "lives on his nerves, thinks he knows better" (*CP* 40).

Bishop seems in Key West to have found solace in the adequacy of the reduced and given, the *everyday* an orientation with obvious application for the Anthropocene. The unpainted pine siding of the dwelling she describes in "Jerónimo's House" resemble a "gray wasp's nest/of chewed up paper/ glued with spit." If so, the fact does not prevent its effective functioning as "fairy-palace," "love-nest," and "shelter from the hurricane." The poem invites us to examine a Key West "conch" neighbor's ramshackle dwelling and to find it and its eccentric furnishings entirely adequate and sufficient. To find them so, to *think of* them this way, becomes the explicitly repeated injunction of "Little Exercise," as well, Bishop's primer for the poetic imagination. With the title's gesture toward the scales and arpeggios commended for beginners, the little exercise is a drill or warm-up for poets and would-be survivors. The exercise can be performed in bed in the middle of the night as a storm passes overhead: "Think of [this as that]," the speaker repeats. Draw upon your stored impressions of the closely observed daylight world and imagine the effects of the storm passing: "Think how they must look now, the mangrove keys/lying out there unresponsive to the lightning in course-fibred families." One of Bishop's gothic early poems in *North & South*, "The Unbeliever," drew upon a figure in Bunyan to imagine its faithless protagonist sleeping "at the top of a mast" in constant danger of falling. "Little Exercise" reverses that image, laying claim to the security of drawing in memory upon close and carefully imagined observation:

> Think of someone sleeping in the bottom of a row-boat
> tied to a mangrove root or the pile of a bridge
> think of him as uninjured, barely disturbed.
>
> *(*CP *41)*

Think of the Keys-freshening storm passing in the night, and of the world as it should be. Think of the body as secure and whole, at ease. Think of human consciousness as afloat in the world's midst, immersed but unprovoked within the scene. Simple though it is, this state of adequacy and continuance, hard-won in Bishop's life, is the anchoring invitation of her best work.

Against the challenges of the Anthropocene, one may well doubt whether the "uninjured, barely disturbed" state of Bishop's sleeper in the boat will go far. To imagine someone sleeping undisturbed through the storm – floated by the rowboat, tied to the mangrove roots – is another *little exercise*, one of surviving without effort, and without even noticing, secure at the heart of the storm, a powerful alternative conception of *Anthropos*. The *little exercise* of blending the anchor of close empirical attention with the animation of conjured imagination provides a clue to "The Fish," another of Bishop's most important Key West poems. The lines begin narrating a process by which the speaker examines a venerable fish hooked and held to the side of her boat. But after 25 lines on the fish's exterior, and as if "driving toward the interior" as another poem puts it,[8] the speaker plunges past the "gills/fresh and crisp with blood" and enters into the fish's imagined body, narrating as she goes: "I thought of the coarse white flesh/packed in like feathers, * * */the dramatic reds and blacks/of his shiny entrails,/and the pink swim-bladder/like a big peony." The impressions conveyed are not remotely flights of poetic fancy but rather a fisher's accumulated knowledge recollected. Precise, accurate, remembered from experience, the details drawn upon and spoken create a space for a palpable sense of the value of experience, precisely the point of the "beard" of swivels and wire leaders that the fish trails from its venerable jaw.

As the fish's imagined viscera convey a sense of the wondrous, so these features make legible a story of survival: a "green line, frayed at the end/where he broke it" and another, "a fine black thread/still crimped from the strain and snap/when it broke and he got away." The speaker translates this "beard" of lethal tackle into a decorated soldier's "medals," an acknowledgment of exemplary service in the postponement of fate. Built of the recognition of such scrutiny, the speaker's regard for the fish accretes, as does her perception of a sense of "victory fil[ling] up/the little rented boat." An enveloping perception of "rainbow, rainbow, rainbow" brings about the concluding action: "I let the fish go" (*CP* 42–44).

How shall we gauge the meaning of this fishing expedition in which the speaker, landing a tremendous fish, investigates the story that the great creature's body tells and, bearing witness to that record of victory, receives a vision of rainbow, and releases the fish? Bishop seems in part to turn here from her Key West neighbor Ernest Hemingway's Gulf Stream pursuit of Holocene-era rituals of human dominance, repeated reenactments of grappling with sea beasts and prevailing over them. She signals the latitude of her rejection of such human traditions of correspondence in part by assigning the thrice-voiced rainbow to the prismatic effect of engine oil floating on bilge water at the bottom of the boat. The delicate rainbow glimpsed is no sign of the God's promise to withhold himself from further destruction (Genesis 8), but a taint all too typical of interactions between Anthropos and the environment. Bishop was herself an enthusiastic and skilled fisherman but she turns self-consciously against the heroics of self-aggrandizement and possession expressed as the very logic of fishing by her neighbor to land a compelling alternative: attending to embodied life, honoring survival, allowing for continuance. Remarkably enough, her poem may have inspired even in Hemingway an envy for realities he tended to overlook. Bishop is said to have prized above any praise the poems of *North & South* received in academic circles the response to "The Fish" of Ernest Hemingway, who wrote of the poem to his ex-wife, Pauline, Bishop's friend in Key West, "I wish I knew as much about it as she does" (Travisano 168).

That phrase, the idea of wishing we knew about it as much as she does, nicely catches what it is like to read Elizabeth Bishop. The speaker's close reading lands by the end of the poem an object more "tremendous" than the "fish" hooked at the start. The idea arrives in this human-animal interaction only through the fisher's rapt attention to the fish, its body, its beauty, its history, its lice, its way, its weight. The sense of "victory" seems built up out of acts of recognition. Iris Murdock observes that "Love is the extremely difficult realization that something other than oneself is real" (215; qtd. by Jamieson and Nadzam 204). Such a difficult realization, the alternative knowledge Hemingway envied, comes to the speaker through a close reading of this heavy, grunting, weighted, real creature. And it comes to the reader as well, who is provoked, instructed, alerted through this poet's grappling not against a venerable sea beast with the intent to subdue it but with a set of automatic habits and too little considered assumptions from the Holocene about human wishes for contest and mastery.

For Stevens and Bishop, gaining more accurate readings of the world about involves us in getting beyond hardwired and conventional views, a voluntary stepping downward and inward toward closer readings. For both, this getting further, though initially a doubt or skepticism, a going without, is richly rewarded with the bliss of seeing more of what is actually there. The work entailed involves the deeply moral "art" of losing such as Bishop extols it in her poem "One Art" (1976), where, with a rueful certainty, she commends the experience of losing as the only game in town for Anthropos. We are perpetually *losing* all our lives, she observes, things banal and momentous, and the only "art" available to us is to respond to the situation with equanimity and accurate perception of our own powerlessness. "[S]o many things seem filled with the intent/to be lost" that their coming to be so is less grounds for identifying the situation as "disaster" and thus coloring it with our own investments, than it is the playing out of a norm. Loss is simply what happens, and quite separate from our ideas about it. What more appropriate skill or practice for the Anthropocene than to "Lose something every day," as Bishop's *art* in this poem and others advises; "Accept the fluster . . ./. . . Then practice losing farther, losing faster." This deadpan emotional advice is tough love for narcissists and solipsists, encouraging us to close read the scene of reality. Contemplating Bishop's *art of losing* in relation to "the hour badly spent" or even one's "mother's watch" is one thing; but wouldn't the present and impending losses of ocean acidification, species collapse, climate change count as "disaster?" Possibly, though that word's root meaning is "ill-starred," or *unfortunate*; the causes driving such changes to the Earth system seem to lie far closer to hand. Bishop's point likely holds.

The good commended in "One Art" will not save the world, nor the creatures in it, including ourselves. But as the globe heats and familiar conditions reel, Bishop's work commends a tough-minded attentiveness and – "(*Write* it!)" – the making conscious of what, as we see, we are losing. Such expenditures of attention could yet transform our own ethics and practices, inspiring better, more limited, more attuned ways of inhabiting this excellent fair planet. They could also transform our self-conceptions, restoring the sense identified and celebrated by Stevens in "Evening without Angels," of the enduring "great interests" of Anthropos: "air and light, the joy of having a body, the voluptuousness of looking" (136).

Notes

1 As Ian Angus notes, "[the term] Anthropocene does not refer to all humans, but to an epoch of global change that would not have occurred in the absence of human activity" (232).

2 ". . . Stevens is the most Coleridgean of the Modernists, most notably in articulating a postromantic theory of the imagination that claims transformative power for the imagination without Coleridge's transcendent metaphysics. What is more, and somewhat more surprising, Coleridge was prominent in Bishop's pantheon as well" (156). "That adorable man!," Bishop called him in a 1956 letter (qtd. by Gelpi, 156).

3 Recent theorists of ecopoetics describe approaches to reading that treat not only correspondences across the gap between "human and nonhuman nature," but also that between "two areas of intellectual endeavor: ecology and literary study." For William Rueckert, ecopoetics approaches reading itself within "the first law of ecology" – that "Everything is connected to everything else." What more could a New Critic say? Citing the "earth-centered" and "environmental" approaches to reading of Cheryll Glotfelty and Lawrence Buell, Selby posits that poetry itself becomes "a structure of active engagement with reading as an environmentally conscious act" (Selby 128–29).

4 Scroggins rightly emphasizes Stevens's response to the lack of spring in Florida as the ground of "Indian River," quoting a letter home to Connecticut in which Stevens extolled "our rich variety of four seasons, our Exquisite Spring and long autumn . . . variety that the lotus-eaters of the South must pine for" (38; Letters 211)

5 Travisano (164), Kalstone (63), and Wojahn (40–41) all describe the shift in *North & South* away from surreal fables expressive of Bishop's anxieties of rootlessness ("Sleeping on the Ceiling," "The Man-Moth," "The Weed," "The Monument") and the rise of a new mode of attention, naming, description that coincides with Bishop's settling in Key West.

6 Travisano notes that Bishop's descriptions "are full of playful hints at the way human observers bend their sense of nature in directions that they find familiar" (165).

7 For discussion of an important debate in Stevens's reception, see "The Everyday: Andrew Epstein's account of Stevens's work as an attempt everywhere to draw close to *the ordinary*.

8 Cf. "Arrival at Santos."

Works cited

Angus, Ian. *Facing the Anthropocene: Fossil Capitalism and the Crisis of the Earth System*. Monthly Review Press, 2016.

Bishop, Elizabeth. *Elizabeth Bishop: Poems, Prose, and Letters*. Robert Giroux and Lloyd Schwartz, eds. Library of America, 2008.

———. *Poems*. Farrar, Straus, Giroux, 2011.

Brooks, Cleanth. "The Language of Paradox." *The Well-Wrought Urn: Studies in the Structure of Poetry*. Harcourt, 1947, pp. 3–21.

Cronon, William. *Uncommon Ground: Toward Reinventing Nature*. Norton, 1995.

Darwin, Charles. *Evolutionary Writings*. James Secord, ed. Oxford UP, 2008.

DuBois, Andrew. "Introduction." *Close Reading: The Reader*. Andrew DuBois and Frank Lentricchia, eds. Duke UP, 2003, pp. 1–40.

Emerson, Ralph Waldo. *Essays and Lectures*. Joel Porte, ed. Library of America, 1983.

Epstein, Andrew. "The Everyday." *Wallace Stevens in Context*. Glen Macleod, ed. Cambridge UP, 2017, pp. 326–34.

Gallop, Jane. "The Historicization of Literary Study and the Fate of Close Reading." *Profession*. Modern Language Association, 2007, pp. 181–86.

Gelpi, Albert. "Wallace Stevens and Elizabeth Bishop at Key West: Ideas of Order." *Wallace Stevens Journal*, vol. 19, no. 2, 1995, pp. 155–65.

Hamilton, Clive. *Defiant Earth: The Fate of Humans in the Anthropocene*. Polity, 2017.

Hofmann, Michael. "The Emperor of Nonsense." *New York Times*. 21 December 1997. Sect. 7, p. 9.

Jamieson, Dale and Bonnie Nadzam. *Love in the Anthropocene*. OR Books, 2015.

Kalstone, David. *Becoming a Poet: Elizabeth Bishop with Marianne Moore and Robert Lowell*. Farrar, Straus, Giroux, 1989.

Latour, Bruno. *Facing Gaia: Eight Lectures on the New Climatic Regime*. Polity, 2017.

Melville, Herman. *Moby-Dick*. Hershel Parker, ed. 3rd ed. Norton, 2018.

Murdoch, Iris. "The Sublime and the Good." *Existentialists and Mystics: Writings on Philosophy and Literature*. Penguin, 1997, pp. 205–20.
Scroggins, Mark. "Florida." *Wallace Stevens in Context*. Glen Macleod, ed. Cambridge UP, 2017, pp. 37–45.
Selby, Nick. "Ecopoetries in America." *Cambridge Companion to American Poetry since 1945*. Jennifer Ashton, ed. Cambridge UP, 2013, pp. 127–42.
Stevens, Wallace. *Collected Poems of Wallace Stevens*. Knopf, 1967.
Travisano, Thomas. *Love Unknown: The Life and Worlds of Elizabeth Bishop*. Viking, 2019.
Wojahn, David. "'The Fiery Event of Every Day': Bishop, Reagan, and the Making of *North & South*." *American Poetry Review*, vol. 41, no. 6, 2012, pp. 35–41.

8

THE TREE AS ARCHIVE

George Nakashima and the nuclear age

Isabel Duarte-Gray

> Peace is what we need and what we all pray for. That is why the High Symbol of the Table as a spiritual crossroad, as a meeting point makes more sense than ever. Mutual understanding between people and generations, parents and children, tribes and confessions is what really matters . . . Today the symbolic and practical message of understanding and conference is what matters so much and I cannot overestimate its role.
>
> (Kapitsa)

I begin with lines from a letter written by Russian public intellectual Sergei Kapitsa to the family of George Nakashima, the "Elder Statesman of the American Craft Movement," concerning Nakashima's offer of a "peace altar" to the Soviet government in 1987 (Ostergard 17–20). Kapitsa, writing well after the fall of the Berlin Wall in 1992, still recognized the symbolic power such a table – be it the Arthurian round table, or Nakashima's signature free-edge slab – around which compacts of peace might be made. What Kapitsa understood implicitly, and what I will argue in this essay, is that Nakashima's pacifist vision coincided with his peculiar modernist esthetic, and both grew in the shadow of the nuclear age, or what Rey Chow calls the "age of the world target" (Chow). Nakashima designed his Peace Altars to gather civilians and representatives of global superpowers to the same table, in service of the same goal: to curb nuclear proliferation. His Altars employ the medium of wood grain to place the natural world at the center of the nuclear discussion, to ask how civic action functions in an era of unprecedented, and even apocalyptic, global interdependence of individuals, nations, and living things. Nakashima's Peace Altars, though quixotic, make a case for the power of civilians to advocate for their own collective self-interest on the global stage in absence of a functional representative democracy and coherent social contracts.

George Nakashima, ecocritic?

Though he was trained as an architect at MIT and the L'École Américaine des Beaux-Arts at Fontainebleau, George Nakashima turned his attention to joinery precisely when the world was split along fronts, alliances, and lines of loyalty, and as the United States was discovering atomic fission. While Nakashima produced substantive work before the United States entered World War II, it was during his 1942–1943 internment at the Japanese concentration camp of Minidoka that he apprenticed informally with the master wood craftsman Gentaro Hikogawa, and immediately after internment that he fully devoted his career to the craft. In order to leave the internment camp with his family and infant daughter, Nakashima formally renounced claim to his property and architectural profession, and accepted work as a chicken farmer in New Hope, Pennsylvania. There, he built a workshop and woodworking studio, which was met with near-immediate commercial success.

In *Full Circle*, a retrospective of Nakashima's work, Derek Ostergard argues that Nakashima's single greatest contribution to modern furniture design is the free edge – the careful, extemporaneous use of natural wood edges and figuration, taken from enormous slabs of oak, walnut, maple, redwood, and so on, such that his furniture resembles both an architectural creation and a living thing (22–23). Given the free edge's legion of imitators, one must look to Nakashima's design context to understand its originality and historical significance – which, per Ostergard, is best viewed in comparison with similar designs of the period, as epitomized by the New York Museum of Modern Art's competition, *Prize Designs for Modern Furniture*. The free edge diverges from American furniture, both classic and midcentury, in that it preserves, as much as possible, the original shape of the tree from which its wood was harvested, as opposed to cutting boards into straight edges,[1] or, as was common in the MoMA exhibition, creating man-made biomorphic shapes influenced by the surrealist forms of Miró, microscopic organisms, and the "melting blobs of atomic warfare" (Koplos 227, 214). Moreover, Nakashima's hand-planed, locally-sourced wooden furniture emerged contemporaneously with mass-producible designs manufactured almost entirely from synthetic materials, of which the most famous is the Eames molded fiberglass chair (Ostergard 68–69). Per the MoMa design catalog, the intent of the exhibition was to "make clear the urgency of the need for furniture that gives better values not only pricewise but also in the efficiency of its planning and in its fine appearance" (Kaufmann 8). Two designs shared the first prize, both mass-manufactured chairs: the first, designed by San Franciscan Don Knorr, was made from sheet metal bent to conform to the human body and coated with a flexible synthetic paint, while the second, designed by Georg Leowald of Berlin, was a plastic, stackable chair with interchangeable parts to reduce shipping and storage space. The catalog noted that the latter chair's design could not be realized "with the limited facilities available in postwar Germany and would require further investigation along lines which have not yet been undertaken in this country," implying that the period's esthetic values grew from the clean lines of late Modernism, the politics of international cooperation in an era of expanding global capitalism, and the problems of (global) post-War austerity (Kaufman 16).

Nakashima would outline the philosophy that precipitated his formal and material divergence from this design moment in *The Soul of a Tree* (1981), but his thinking on this subject had already reached a sophisticated degree of articulation by 1953. Three years after the seminal MoMa exhibition, Nakashima addressed a New York *Herald Tribune* forum on contemporary design, in a speech entitled "One Man's Answer," which helped cement him as one of the most important practicing midcentury designers in the United States (McQuaid). Nakashima's "Answer" speaks to his relationship with the Anthropocene, by way of his relationship to raw material. The question he implicitly "answers" emerges organically: how does the artist manufacture the objects of daily life, and with what ingredients does he create a better life? He writes:

> Fundamentally, I have a primary interest in natural materials: the deep wearing of centuries of weather on the timbers of Horiuji, the clean freshly hand-planed surfaces of Hinoki or Port Orford cedar, the burls of English oak, the textures, colors, and the sheer art and craftsmanship of a good fieldstone wall, the stripped beauty of high-strength concrete, the character and history of a wide plank of walnut, the life and death of two hundred years, the story of bitter winters and the searing desert summers of at least a human lifetime that show in the twisted, tortured forms of bitter brush, the story of a beautiful wide unvarnished boards of a teahouse ceiling: these are the fine relationships of man with nature, the relationship of man with a higher being; his universality with his environment, the possibility to embrace it instead of conquering it. What can plastic offer? Or fiberglass? In these uncertain times, this type of norm is important.
>
> *("Answer" 59).*

This passage contains, in its principles of selection, both Nakashima's response and his condensed autobiography, told through his observation of natural objects across time. The Buddhist temple of Hōryū-ji is one of the oldest standing wooden buildings on earth, with timbers dated back to the sixth century ad. His love of Hinoki and Orford cedar alludes to his admiration for Japanese woodworking techniques and materials, but also to his childhood and early woodworking period on the Pacific West Coast, where Port Orford cedar is native. The fieldstone walls and English oaks are endemic to his adoptive home of Bucks County, Pennsylvania, while the concrete refers equally to his newly poured concrete home in New Hope, which he designed from scratch, and to his formative experiences in Paris, watching French workers pour the foundation to Le Corbusier's Swiss Pavilion in 1933 (Mira Nakashima 18).

The most telling choice, however, in this list of observations is the selection of bitterbrush. This shrub, which Nakashima invariably calls "bitterbrush" and not by its more common name, greasewood, is a species he encountered during walks in the Idaho desert during his internment at Minidoka. In *The Soul of a Tree*, Nakashima writes that he and Gentaro Hikogawa collected and polished samples of bitterbrush ("a brave shrub of great character which grows only a few feet in a hundred years,") to give away as gifts and decorate their living quarters in the camp (*Soul* 70). The "bitter winters and searing desert summers" of the Idaho desert move a step beyond the anthropomorphization of a brave little shrub: they liken Nakashima's suffering in a crudely constructed and frequently flooded concentration camp to the stalwart endurance of the native Idaho flora. Nakashima's encomium to natural material, here, is at once the story of his life, told through the observation of beauty, and an appreciation of natural wonders that dovetail with his life, extended beyond the scope of human time.

The timescale of his chosen artifacts is wide, but distinctly not anthropocentric: it includes the timbers of a 1,500-year-old Buddhist temple, a centuries-old American wall made from the refuse of geologic time, and the biography of a being that lived for centuries, told through the careful finishing of book-matched timber. Nakashima's "answer" to an unspoken question is to seek an art form that transcends its era; to draw out, with humility, the beauty that exists already within extant material, rather than imposing a human vision *upon* raw material.

To close read a tree

I have posited that Nakashima came to his esthetic signature during the age of the world target. Rey Chow's thesis of the "world target" draws from Heidegger and has two chief elements. First, the dropping of atomic bombs on Hiroshima and Nagasaki affected the majority of the world population as a visual image of terror, which Chow calls "an epistemic event in a global culture in which everything has become . . . visual representation and virtual reality" (Chow 26–27). The fear of deterrent weapons and their detonation now infiltrates the mundane world and hovers over civilians like pall. Hence Chow's second point: the indefinite state of potentiality that waits in nuclear weapons removes the line between a battle front and civilian space: "The mushroom cloud . . . is . . . the image of this semiotic transfer, this blurring of the boundary between war and peace" (Chow 31–32). Historian Peter Galison has illustrated this principle compellingly in "War Against the Center," in which he demonstrates how the U.S. federal government launched a totalizing policy of industrial dispersion in 1946 to decentralize urban planning as much as possible, to create fewer and smaller targets for aerial bombardment (13–15). Galison concludes that the U.S. government's inquiry into the damage it inflicted with aerial bombs created a new kind of "atomic imaginary" joined to the most mundane acts of urban planning, from phone lines to electrical grids:

> Here stands a new, bizarre, and yet pervasive species of Lacanian mirroring. Having gone through the bomb-planning and bomb-evaluating

> process so many times for enemy maps of Schweinfurt, Leuna, Berlin, Hamburg, Hiroshima, Tokyo, and Nagasaki, now the familiar maps of Gary, Pittsburgh, New York City, Chicago, and Wichita began to look like them.
>
> *(30)*

While it is hardly novel to observe that the "atomic imaginary" is a fundamental element of post-45 American thought, balancing the apocalyptic and the banal in manic tension, that same tension contextualizes Nakashima's naturalistic esthetics.

What Chow calls the "semiotic event" of the nuclear bomb, which "blurs" all previous boundaries between war and peacetime, Giorgio Agamben might call the growing state of exception among democratic states, the "uncertain and ambiguous zone in which de facto proceedings, which are themselves extra- or anti-juridical, pass over into law, and juridical norms *blur* with mere fact – that is, a threshold where fact and law seem to become undecidable" (Agamben 30, emphasis mine). Beginning roughly with Hitler's indefinite suspension of the Weimar constitution, the "logic and praxis" of juridical, constitutionally articulated, or otherwise socially contracted rights have blurred and "a pure violence without logos claims to realize an enunciation without any real reference" (Agamben 40). In other words, modern states have incorporated "pure violence" or violation of natural rights into the "blurry" space of the legal – as in the Third Reich's suspension of the Weimar constitution, the mass deportation and eventual incarceration and slaughter of Jewish citizens in Germany, the mass slaughter of Japanese civilians in Hiroshima and Nagasaki, the indefinite incarceration of Japanese American civilians without charge or trial, and any number of contemporary state-sanctioned violations of civilian rights. This "blurriness" – both of the social contract and of the line between the civilian and the military target – requires us as scholars and citizens to reconsider the question of the legible: What rights are legible, and what beings are afforded legible rights, under the modern social contract? How do civilians begin to articulate such rights under increasingly "ambiguous and uncertain" conditions? I will turn, as it were, to one man's answer.

Nakashima's slab tables, which his Altars for Peace epitomize, are distinguishable by two features: the free edge and the "butterfly key," a visible form of joinery. Both features help realize his structural goal of appreciating the integrity of an ancient tree as it stood, and both play a role in the metaphorical content of his works, as they remind us of the brevity of human lives and of the fragility of the environments in which we – humans, animals, and trees – exist.

Nakashima writes that "there is drama in the opening of a log – to uncover for the first time the beauty in the bole, or trunk, of a tree hidden for centuries, waiting to be given this second life" (*Soul* 95). He developed meticulous harvesting methods to preserve ancient trees as completely as possible, because he believed there was one ideal use to feature a tree's grain and shape – to honor the history of each tree. The single walnut tree, which made up Nakashima's Altars for Peace, took *days* to cut into individual pieces, and included the discovery of a piece of pipe at the center of the tree. The *National Geographic* documentarian who witnessed this process described it as a matter of war and peace intervening in the history of flora: "Trees cannot hide from men. Instead they stoically submit to and absorb the fleeting presence of each human generation they outlive. Anchoring hooks from forgotten clotheslines, bullets meant to kill men in war all become part of a hidden history beneath the bark" ("Elegant"). Nakashima's sense of the human life, entwined with the nonhuman lives, was heightened in the wake of World War II, the Vietnam War, and the Cold War: his free edges tell the story of lives outside the global public sphere, opened and book-matched so the grain is legible to the human eye.

Nakashima's signature form of exposed joinery, an ancient device called the "butterfly key," is a small, hand-cut piece in the shape of an hourglass or butterfly that allows cracked wood (cracks, radiating from the center of the tree often form in the attempt to harvest full slabs) to be handled as though it were whole. Just as the Shakers, from whom Nakashima partly borrowed this device, treated their work techniques as both literal functions and metaphorical tools of self-discipline, Nakashima's solution to a practical problem has metaphorical power: Nakashima leaves marks of human intervention upon the natural surface of wood instrumentally, to strengthen weaknesses and to join points of divergence. Like Shakers, Nakashima's use of joinery is performatively visible, an act of gathering together that which threatens to fall apart, and performatively fragile, mimetic of the most filmy and ephemeral of living organisms, the butterfly.

FIGURE 8.1 Butterfly keys, clearly visible as George Nakashima inspects the Altar for Peace at the Cathedral of St. John the Divine.

Source: Courtesy of the Nakashima Foundation for Peace

Altars for peace: the conception (1969–1983)

While Nakashima's esthetics are, I have argued, reactive to the age of the world target, his most explicitly political work, the Altars for Peace, occurred as a result of several coincident events in his life. The first of these was George Nakashima's growing friendship with Harvard scientist and Nobel Laureate George Wald. The two met in the late 1950s, after Wald attended a lecture by Nakashima and was captivated by the lyric quality of his designs (Wald Letter 1979). Wald would later write of his first impressions of Nakashima's work that "his is an art of subtraction, of removing the obstructions, the barriers, to let the inner meaning emerge; as in Michelangelo's uncompleted 'Slaves' in Florence, where you see it happening" (Wald Letter 1979). The two struck up an epistolary friendship, followed by a series of family visits that would continue until Nakashima's death in 1990. Their correspondence became particularly dense, and vital to each, as they realized their mutual opposition to U.S. foreign military intervention and to the American policy nuclear deterrence.

This connection materialized as a result of George Wald's 1967 Nobel Prize, or rather, as a result of Wald's decision to use his Nobel publicity to oppose the Vietnam War. In 1969, Wald spoke at MIT, addressing the unrest of the first generation raised in the age of the world target. His students, he argued, seemed "beset with a profound uneasiness," which he attributed to the War and to the apocalyptic logic of the "atomic imaginary" ("Generation" 351–56). Student unrest occurred, Wald said, because "mutually assured destruction" denied them the human right to an assured future, and thus, of the right to strive for that future (356). Wald concludes his argument by sketching out the interlocking human and nonhuman stakes of nuclear disarmament, in a thought exercise that resembles Nakashima's own philosophy:

> I tell my students, with a feeling of pride that I hope they will share, that the carbon, oxygen, and nitrogen that make up 99% of our living substances were cooked in the deep interiors of earlier generations of dying stars. Gathered up from the ends of the universe, over billions of years, eventually they came to form in part the substance of our sun, its planets, and ourselves. Three billion years ago, life arose upon the Earth. . . . About two million years ago, man appeared. He has become the dominant species on Earth. All other living things, animal and plant, live by his sufferance.
> *(356)*

Wald's appeal to humility in the face of nonhuman temporalities, and the deep interconnectedness of "living substances," recalls Nakashima's philosophical sensibility – his fieldstone walls and Horiuji timbers.

Soon, both Nakashima and Wald would read Jonathan Schell's best-selling essay *The Fate of the Earth*, a sober projection of the global aftermath of nuclear war, grounded in contemporary scientific data. The final section of Schell's essay makes a few prescient charges; like Wald, Schell argues that the mere existence of nuclear arsenals chokes off the future – and therefore present – of humanity; like Rey Chow, he observes that the brinksmanship of nuclear deterrence erases the distinction between wartime and peacetime, battle fronts, and civilian spaces. Finally, like Agamben, Schell's essay addresses the question to which Nakashima will turn in his Altars for Peace: the problem of the social contract, of *consent*, in an age of mutually assured destruction and global warfare. Schell argues, borrowing from Hobbes, that nuclear deterrence – as an implicitly global phenomenon – exists in a constant suspension between the "state of nature," in which individuals live without a centralized governing authority, and a "civil state," in which individuals consensually surrender the capacity for violence to a government, which uses that capacity for the common good. Schell writes that the global superpowers' policy of nuclear deterrence forfeits individuals' capacity for self-defense at the global level without providing any authoritative governing body or common interest (215–16). Thus, at the global level, human beings are not granted the possibility to consent or dissent with regard to their total annihilation, *though that annihilation can still take place*. Or as Nakashima wrote to Wald: "I still have faith that it is not man's prerogative to wipe himself off the face of the earth, although I am sure he has developed the capacity to do so" (Nakashima Letter 1982).

The following year, Nakashima first described his "dream" to Wald – and later, the New Hope *Gazette* – after purchasing the raw materials for what would become his Peace Altars. His vision appeared, he wrote, after undergoing anesthesia for an operation, though as we know (and as he told the New Hope *Gazette*), the rumblings of an artistic statement urging peace – and opposing nuclear arms – had been sounding for decades. Nakashima writes of a utopian construction, of specific design and unspecified location:

> There has appeared an extraordinary natural phenomenon, something that occurs only once in a lifetime or perhaps only once in the history of a nation or in all time – It is a great walnut tree. It is a tree that could be a symbol of nature and man in the deepest spiritual sense. It is now on hand. . . . A pilgrim shrine, a temple, a great half sphere rising out of the water, perhaps a hundred meters wide and fifty meters high, sheathed with fish scale like ceramic tiles in white and pale yellow to glow as a symbol in a stretch of water to make a complete sphere. It might be on land: or the East River, or the San Francisco Bay, or the Tokyo Bay, or on the Ganges, somewhere where the sadhus could pay homage to an ideal, an ideal to which they have dedicated their whole lives.
>
> *(Nakashima Letter 1983)*

The vision begins with the third condition that generated Nakashima's eventual Peace Altars; the discovery of a walnut tree older than the United States itself. Nakashima outlines his hazy intention to place its heartwood at the center of a monument, meant to serve as a beacon of calm and humility in the space where some colossus would normally stand. The vision ends, however, with a clear articulation of why his walnut tree should be used as a symbol of peace, and what specific historical arrival it reacts against:

> Perhaps this is all a dream, but peace is a dream, a dream dreamt by an overwhelming mass of millions upon millions of people. It cannot be had within the halls of power as ego begets ego until we have our final Armageddon. . . . Peace should be born as a genuine expression of nature and an act of beauty. There can be at least one small spot on earth to be dedicated to Peace in a tangible form instead of an abstract idea and an abuse of war. It can be a positive creative force of its own carrying its own momentum.
>
> *(Nakashima Letter 1983)*

Nakashima's belief that "an overwhelming mass" of people yearn for protections from harm, which desires go unheard by the extraordinarily small number of representatives in power, echoes Schell and Agamben, and calls to mind the work of Elaine Scarry on consent and "out-of-ratio" weaponry. Scarry, like Schell, believes that "out-of-ratio" weapons – "any form of weapon that allows a tiny number of people to kill many millions of people" – violates the fundamental tenets of government as a social contract, in that it deprives the injured civilians of the right to self-defense and the *offending country* of the right to self-defense, in that they are eliminated from both the decision-making process (through both direct dissent and parliamentary or legislative consent) and the act of defending themselves directly (4–6). In other words, Scarry elaborates, if the social contract is fundamentally a "compact for peace," designed to prevent injury against the human body, a technology that puts massive killing power into the hands of very few deprives the injured and injuring populations of the right to consent to acts of war-making (Scarry 5). Scarry's conclusion arrives where Nakashima's Peace Altars begin, with raw material of trees:

> Historian Peter Linebaugh has made us aware that the Magna Carta originally contained another charter, which then grew up side by side with it. It was called the Charter of the Forest. It sought to guarantee that the bounty of earth was available to all people, not just to kings, as had been true earlier. So, too, the obligation to protect the earth belonged to all people, not just to kings. . . . The forest's 'greenhue' or 'vert,' as Edward Coke explained in his seventeenth-century margin glosses to the 1224 document, was 'whatsoever beareth green Leaf.' It was available to the population for diverse uses such as shelter ("Haw-Thorn, Black-Thorn") and food ("as Pear-Trees, Chestnut-Trees, Apple Trees"). Included among these uses – explicitly and repeatedly – was self-defense: Oakes, Beeches, Ashes, Poplers, Maples, Alder, and Elder. The charter addressed not just

FIGURE 8.2 Nakashima displays the Altar for Peace in his workshop in New Hope, Pennsylvania.

Source: Courtesy of the Nakashima Foundation for Peace

> the need to distribute the resources of earth away from kings and toward the people but also the need to distribute them across generations.
>
> *(398–99)*

The materials that were once used for war-making, for battle rams and longbows, may also serve to remind us that the stakes of split-second nuclear decisions are not only measured in existing human lives, or in the lives of beings who can offer their consent, but also in the possibility of sustaining history, or any sense of our place in time: in the careful preservation of history, and thus, the preservation of life itself.

Altars for peace: the development (1984–1989)

The making of the Altars for Peace began with cutting of the walnut tree into slabs, which would be dried and finished over a period of years. In early 1984, Nakashima – anticipating the massive fundraising effort to come – turned the act of harvest into a public event, inviting both his local newspaper and a crew from *National Geographic* that had already filmed a piece on his work to witness the multiday process. Nakashima's walnut tree was, indeed, unique; it was so large that Nakashima was obliged to hire a professional sawyer who specialized in walnut from California to fly East with a custom saw commissioned for the purpose (Gazette 3,14). Both the New Hope *Gazette* feature and the *National Geographic* piece highlighted Nakashima's nascent Peace Altar plan and its ideology; Nakashima furnished the *Gazette* with his entire "Altar to Peace" letter, which he had sent to Wald and a number of confidantes the year before, to publicize his vision. Nakashima later wrote to Wald that the *National Geographic* piece, which aired nationally, "might promote something of a more serious nature" – namely, *plural* altars to peace, located at international sites of war-making and reconciliation (Nakashima Letter April 1985).

This vision began to feel more accessible once a motley assortment of Nakashima's former patrons and friends – including members of the Rockefeller family, for whom Nakashima had performed a number of major furniture commissions – formed the Altar for Peace Foundation in 1985. The Foundation put Nakashima immediately in contact with the Church of St. John the Divine in New York City, where the first Altar to Peace would be christened, and empowered him to think of the Altar project as a matter not of civilian meditation, but of civilian diplomacy. By March, the Foundation had begun to seek envoys to make contact with Japan and the USSR to place altars in Hiroshima and Moscow (Nakashima Letter March 1985).

The dedication ceremony, which took place on New Year's Eve of 1986, was both a free ceremony and a social fête: 5,000 people attended the dedication ceremony at the Cathedral of St. John the Divine, with a free concert conducted by Leonard Bernstein. Nakashima saw the dedication ceremony as an opportunity to build the "momentum" of peace he describes from his initial "dream." He explained to the crowd that he was negotiating an Altar in Nagasaki and possibly the Soviet Union: "Peace," he announced, "is much more than an absence of war. It is a creative spirit, a great light that can simultaneously take over the world" (Shaw).

FIGURE 8.3 An overhead view of the first Altar for Peace at the Cathedral of St. John the Divine.

Source: Courtesy of the Nakashima Foundation for Peace

The rousing success of Nakashima's first Altar left Nakashima with great hopes for his most ambitious international project, the Moscow Altar for Peace. His hopes were bolstered by a new professional connection with the Soviet Union; in 1985, George Wald was selected to present a statement drawn up by a committee of Nobel Laureates to Soviet leader Mikhail Gorbachev in advance of his November Summit with President Reagan in Geneva. In 1985, six days before the Geneva Summit, Wald met Gorbachev in Moscow, where he delivered a five-point proposal on part of Nobel Laureates, which included a mutual pledge of no first use, a comprehensive test ban treaty, a massive reduction of nuclear stockpiles, the demilitarization of outer space, and the establishment of a joint framework to prevent the accidental incitement of nuclear war ("Message").

This encounter informed both Nakashima's and Wald's approaches to nuclear disarmament and diplomacy; the latter had already acted in the role of a civilian, on behalf of the powerful Nobel Committee as a diplomat to one of the two most important political agents on the planet. Nakashima felt his art could do the same, speaking not on behalf of an institution, but rather for civilians as a whole, who – rightfully or no – have no power to communicate directly with the possessors of the world's nuclear footballs. Wald's outreach began with letters to two successive presidents of the Soviet Peace Committee, the first of which he had met at a conference the year before. To Yuri Zhukov, Wald and Nakashima offered, at once bravely and presumptuously, "a gift from the American to the Soviet people dedicated to peace" (Wald Letter April 1987). Later, after realizing the Soviet state's official atheism, they clarified to Genrikh Borovik that "this 'altar' carries no particular religious connotation. It is Nakashima's superb expression of his deep concern for life and for peace" (Wald Letter July 1987).

Meanwhile, Wald reached out directly to Gorbachev, whom he had come to consider a friend. Writing of the "life and beauty that only Nakashima can bring out in wood," Wald suggested that the table was "closely related to fostering peace and better relations between our countries," as though the altar's materiality added a more substantial layer to the ethereal official diplomatic relations of the Geneva Summit (Wald Letter August 1987). A few weeks letter, Wald wrote again to Gorbachev, renewing Nakashima's offer, with an addendum as to the use of the now secularized "table": "It occurs to me that this might be placed in the Kremlin as a Table of State, reserved for quite special occasions – negotiations, the signing of agreements – related to peace and international understanding. Its history could grow with its use" (Wald Letter August 1987). Wald's offer recalls Schell in the introduction on *The Fate of the Earth*, quoting President Eisenhower that both nuclear powers would have to "meet at the conference table with the understanding that the era of armaments has ended, and the human race must conform its actions to this truth or die" (Schell 6). While Eisenhower, who also summited with Soviets in Geneva, employs the table as a diplomatic metaphor, Wald and Nakashima consider it phenomenologically, as a temporal one: it is a metaphor for time scale and the stakes of living things in the age of the world target. By placing a table that recalls an ancient living organism literally beneath the process of peace negotiations – as Wald puts it, "a great table . . . the border of which would still follow in large part the outline of the tree" – Nakashima and Wald hoped to remind the negotiators, at every moment of discussion, of the living beings and future lives they are gambling (Wald Letter August 1987). Effectively, the entreaty Wald and Nakashima are making is an act of diplomacy not from one nation to another, but from civilians to their own governments, to remind those governments that living bodies are the material substrate of the social contract.

Final altars for peace (1990–present)

Kremlin officials never responded, and Nakashima's health steadily declined. By 1988, when Nakashima finally made positive contact with the institution that was eventually to receive his Moscow "table" – the Soviet (now Moscow) Academy of Art – he began to doubt that he could physically execute the task of finishing such an enormous table, having suffered a stroke (Nakashima Letter 1988). His final letter to Wald, sent two weeks before his death in 1990, was optimistic, however: thanks to the intervention of an arts advocate and fluent Russian speaker Irene Goldman, the Soviet government had approved the Table for Peace and had invited Nakashima, as an American artistic dignitary, to visit the site where it would be placed. "This I would enjoy," Nakashima writes, "but there is a question of time, as I am not too well and bumble along the way I am doing. . . . I really don't know how much creative energy. . . . I can draw on, though I hope to draw on it as much as possible" (Nakashima Letter 1990). Nakashima died in June of 1990, a year after the fall of the Berlin Wall and a year before the official dissolution of the Soviet Union. The Cold War had ended, and with it, the immediate threat of nuclear war.

In 1992, George Nakashima's daughter, Mira Nakashima, received the letter from Sergei Kapitsa, which I quoted at length at the beginning of this essay. The second Altar was dedicated at the Hague Appeal for Peace in 1999 and installed at the Moscow Academy of Art in 2001. Mira Nakashima eventually dedicated a third Altar in Auroville, India, where her father had designed and lived in an Ashram with the followers of Sri Aurobindo (Mira Nakashima 240–42). The Altar for Peace Foundation (subsequently renamed the Nakashima Foundation for Peace), still headed by Mira Nakashima, hopes to one day place a fourth Altar in Cape Town, South Africa.

I want to return to the words of Sergei Kapitsa on Nakashima's Altars, which ring at present both practical and impossibly quixotic: "Today the symbolic and practical message of . . . conference is what matters so much and I cannot overestimate its role" (Kapitsa). Nakashima's life ended at a time of paradox. The Cold War had subsided, and our collective sense of urgency waned as the perception of imminent nuclear deployment waned. But the age of the world target did not end, nor have the United States and the various possessors of nuclear weapons ceased to stockpile, update, and, just as importantly, *fail to maintain*, nuclear arsenals. The use of these weapons remains contingent on the temperament of a scattering of world leaders and their allegiance to social contracts that are, by the very power of nuclear arms and the epistemological frameworks by which they are controlled, neither enforceable nor coherent.

George Nakashima's Peace Altars remind us of the relationship between art – as a question of human innovation, and of raw, material fact – and these social contracts. Nakashima's first Altar for Peace registers dissent in a time when state use of weapons technology has dissolved the line between civilian and combatant, a line designed to be drawn with public knowledge and consent. It does so by drawing our attention to the patterns figuring across the surface of a being that lived longer than either Cold War nation at that time. Those patterns, the ribbing of single years, lean and plenty, should look like the figuring of history. They should look, too, like an alien thing outside human history, and remind us of the smallness of that history.

Nakashima's second Altar or "Table" for Peace, and his unsent Altar to Hiroshima or Nagasaki, hint at an even larger act of dissent: To send an Altar to another nation, on behalf of Americans, as an act of peace-making or remorse, is to speak words the state officially cannot. To offer a "Table for Peace" to the Kremlin, on behalf of the American people, on which to conduct peaceful negotiations, is to suggest that American elected representatives cannot and do not speak for American interests in matters of nuclear policy. It suggests that artists and civilians must speak for ourselves as civic actors; Nakashima chose to speak through the raw material produced on American soil before it became American soil, through the material weight of the table itself and the material consequences it calls to mind. Nakashima's joinery binds considerations of human agency in the act of making art with problems of civic agency in the act of making war. Nakashima's Peace Altars are not simply the magnum opus of a master craftsman; they address the problem of consent in the age of the world target as a material consideration, as sensate flesh, living things, and objects moving across contingent borders, invariably interconnected and still at risk.

Note

1 Nakashima felt the cutting of large, ancient trees with figuring representative of their age and history into regular, straight boards was a "waste of a majestic opportunity" (*Soul* 93).

Works cited

Agamben, Giorgio. *State of Exception*. University of Chicago Press, 2005.

Chow, Rey. *The Age of the World Target: Self-Referentiality in War, Theory, and Comparative Work*. Duke University Press, 2006.

"Elegant Craftsman." National Geographic Explorer. Nickelodeon. 28 April 1985.

Galison, Peter. "War against the Center." *Grey Room*, vol. 1, no. 4, 2001, 6–33.

Kapitsa, Sergei. Letter to Mira Nakashima. 29 February 1992. Wald, George. Papers of George Wald, 1927–1997. HUGFP 143, Harvard University Archives. Box 14. "Nakashima's Peace Altar."

Kaufmann, Edgar. *Prize Designs for Modern Furniture from the International Competition for Low-Cost Furniture Design*. 1950.
Koplos, Janet, and Bruce Metcalf. *Makers: A History of American Studio Craft*. The University of North Carolina Press, 2010.
McQuaid, Matilda. "George Nakashima and the Mass Production of Craftsmanship." *George Nakashima and the Modernist Moment*. McQuaid, Matilda, ed. James A. Michener Art Museum, 2001.
Nakashima, George. Letter to George Wald. 12 April 1982. Papers of George Wald, 1927–1997. HUGFP 143, Harvard University Archives. Box 14. "Nakashima, George."
———. Letter to George Wald. 9 August 1983. Papers of George Wald, 1927–1997. HUGFP 143, Harvard University Archives. Box 14. "Nakashima, George."
———. Letter to George Wald. 8 March 1985. Papers of George Wald, 1927–1997. HUGFP 143, Harvard University Archives. Box 14. "Nakashima, George."
———. Letter to George Wald. 28 April 1985. Papers of George Wald, 1927–1997. HUGFP 143, Harvard University Archives. Box 14. "Nakashima, George."
———. Letter to George Wald. 13 May 1988. Papers of George Wald, 1927–1997. HUGFP 143, Harvard University Archives. Box 14. "Nakashima, George."
———. "One Man's Answer." New York Herald Tribune. New Patterns for Mid-Century Living: Report of the 22nd Annual Forum [New York], October 18–20, 1953.
———. *The Soul of a Tree: A Woodworker's Reflections*. 1st ed., Kodansha International, 1981.
Nakashima, Mira, and Nakashima, George. *Nature, Form, and Spirit: The Life and Legacy of George Nakashima*. H.N. Abrams, 2003, 18.
"Nakashima Finds Perfect Tree for a Shrine to Peace." *New Hope Gazette*. 9 February 1984. 3, 14.
Ostergard, Derek E., et al. *George Nakashima: Full Circle*. 1st ed. Weidenfeld & Nicolson, 1989.
Photos of Altar for Peace. Courtesy of the Nakashima Foundation for Peace. www.nakashimapeacefoundation.org
Scarry, Elaine. *Thermonuclear Monarchy: Choosing between Democracy and Doom*. 1st ed. W.W. Norton & Company, 2014.
Shaw, Charles. "The 'Dream' is Shared." *New Hope Gazette*. 8 January 1987.
Schell, Jonathan. *The Fate of the Earth*. Avon, 1982.
Wald, George. "A Generation in Search of a Future." *Religious Education*, vol. 64, no. 5, 1969, 351–56.
———. Letter to Dean Norman J. Johnson. 26 October 1979. Papers of George Wald, 1927–1997. HUGFP 143, Harvard University Archives. Box 14. Folder Titled "Nakashima's Peace Altars."
———. Letter to Genrikh Borovik. 29 July 1987. Papers of George Wald, 1927–1997. HUGFP 143, Harvard University Archives. Box 14. "Nakashima's Peace Altar."
———. Letter to Mikhail Gorbachev. 6 August 1987. Papers of George Wald, 1927–1997. HUGFP 143, Harvard University Archives. Box 14. "Nakashima's Peace Altar."
———. Letter to Mikhail Gorbachev. 23 August 1987. Papers of George Wald, 1927–1997. HUGFP 143, Harvard University Archives. Box 14. "Nakashima's Peace Altar."
———. Letter to Yuri Zhukov. 4 April 1987. Papers of George Wald, 1927–1997. HUGFP 143, Harvard University Archives. HUGFP 143, Harvard University Archives. Box 14. "Nakashima's Peace Altar."
———. "A Message to Gorbachev and Reagan." Papers of George Wald, 1927–1997. HUGFP 143, Harvard University Archives. Box 16. "Nuclear Statement Letters."

9

GOING UNDERGROUND

In defense of deep reading

Graham Huggan

What might it mean to "close read" the Anthropocene? The short and inevitably inadequate answer is that it means markedly different things in different contexts. There is no single context surrounding the Anthropocene, still less an "original" context, and no critical consensus on what the Anthropocene is in any case, or on when it might be said to start. To close read the Anthropocene requires the precarious bringing together of what the American literary critic D.A. Miller calls two "cloven" concepts (Miller, quoted in Rooney 137). For Miller, close reading is by definition split between the desire to master the text, effectively to "overwrite" it, and the desire to identify with the text, to respect it by following it as closely as one can (137). Most theories of close reading tacitly accept this double move, whereby the practice of close reading both registers attentiveness to the text and its linguistic strategies, and supplies the foundation for the reader/critic's interpretive claims. Close reading, in this sense, folds (at times uneasily) into *critique*, which is based in turn on two not unreasonable assumptions: first, that it is impossible to describe the text without also interpreting it; but second, that there is no such thing as a "correct" interpretation, turning literary criticism into a politically motivated (but also ethically responsible) play between competing interpretive claims. As will soon be seen, however, the identification of close reading with critique can be challenged, and one of the purposes of this chapter will be to look at some of those challenges, not necessarily with a view to dismissing them, but with a firm push against those scholars, Bruno Latour prominent among them, who claim either that critique has become all too predictable, ossified in its operational maneuvers, or that it has simply had its day (Latour; see also Felski, *Limits*).

The Anthropocene, too, is nothing if not a cloven concept. Even if we accept, as seems reasonable enough to do, that there are multiple versions of the Anthropocene, and that the legitimacy of the claims we make are based on which particular version we are working with, the Anthropocene registers the *limits* of the human as much as it acknowledges the *influence* of the human; it also connotes both *distance* (e.g., the insertion of human history into immeasurably longer earth history) and *proximity* (e.g., the entanglement of our earthly existences with those of others, and the reworking of Harvey's "time-space compression" in a planetary rather than global context that renders us accountable to human and nonhuman lives alike). Unsurprisingly, Anthropocene discourse is fraught with paradox. One of the most notable of a series of possible paradoxes is that *the Anthropocene is not anthropocentric.* Jeremy Davies expresses this particular paradox well: "The epoch does not get its name because nature is now completely subordinated to human agency, as if clouds now form and swallows now fly only after getting permission from human beings" (7). Rather, any meaningful critical engagement with the Anthropocene requires us to accept the ecological truism that: "Humanity is not at the center of the picture of the Anthropocene, oppressing, by its powers of mind, the passive matter that encircles it. Instead, human societies are themselves constructed from a web of relationships between human beings, nonhuman animals, plants, metals, and so on" (7).

A further paradox obtains in this context, namely that *close reading is not close.* This formulation is not as cryptic as it seems. Nor is it restricted to the conditions of the Anthropocene, a reading of whose "scale effects" (Clark, *Ecocriticism*) demands an appreciation of distance as well as proximity: an understanding of the multiple connections, across both time and space, which are needed to make sense of the text. Rather, close reading of whatever kind, which requires a detailed engagement with the text in order to account for the specific rhetorical strategies it deploys as well as the inherent slipperiness of the language it uses, also arguably requires a degree of critical distance. I say "arguably" because this distance can be measured in several different ways, each of which is open to challenge. A certain amount of distance is not just needed in order to exercise critical thought, of whatever theoretical persuasion, but also in order to satisfy the ethical demands of attending to the text. However, as anyone familiar with the recent "method wars" in the United States can attest, this orthodox view of close reading cannot be taken for granted. Take Heather Love, who argues that close reading needs to be unanchored from its "humanist underpinnings," which are linked in turn to a hermeneutic tradition, grounded in "recognition and empathy," that tends to associate the richness and variety of literary expression with the richness and variety of human life (388). Love proposes an alternative model of close reading that draws on the observation-based social sciences to produce descriptions that *are* rich, but not "because they truck with imponderables like human experience or human nature"; rather because they are "close, not deep" and, without "adding anything 'extra' to the description, [are able to] account for the real variety that is already there" (377). This is close reading without critique, attuned to the surfaces of texts and the associative connections traced across them, and suspicious in turn of those kinds of suspicious (symptomatic) readings that assume that texts mean a great deal more than they are prepared to say.

The ongoing, not always especially enlightening debate about the respective merits of "surface" and "symptomatic" reading is familiar enough not to require detailed rehearsal here. Probably also familiar is the connection between this debate and the revisiting of critique as the default mode by which literary and other cultural critics assume – in Elizabeth Anker and Rita Felski's words – that "the essential meaning of a text resides in a repressed or unconscious content that requires excavation by the critic": that requires one form or another of deep reading that recognizes the multiple levels on which literary/cultural texts operate, only some of which are immediately apparent in the texts themselves (16). There's certainly something to be said for the denaturalization of critique, just as critique itself tends to be suspicious of anything that seems to be natural: critique is as deserving of an appropriately critical close reading as anything else. And there's something to be said as well for the questioning of modes of critical reading that assume basic differences between the "surfaces" and "depths" of literary texts. Toril Moi is particularly tough on what she sees as the illusion of deep reading: "Whether they speak of surfaces or depths, readers do pretty much the same sort of thing, regardless of their understanding of what they're doing"; this "pretty much the same sort of thing," though she doesn't say it, is close reading, the main purpose of which is to pay "maximal attention to the words on the page" (35). Close reading, Moi implies, isn't about "describing" (surface reading) or "uncovering" (symptomatic reading); it's about *looking* and *thinking*, which we are all perfectly capable of doing as long as we give due attention to the text (42).

My own view is rather different. "Looking" and "thinking" are, after all, hardly neutral activities, and all critical readings, whether inclined toward surfaces or depths, impose their own interpretive standards, whether these are acknowledged or not. (To be fair, Moi would probably agree with this, her main point being that readers need to *own up* to these interpretive standards: to take responsibility for their own reading practices, and to be wary of the critical doxa that implies that is necessarily better to pick fault with literary texts.) Critique, I would have thought, is as likely to pick fault with itself as with the text, which is one of the many reasons it is valuable, while the received idea that critique is basically extractive – that it relies on prescriptive strategies of deep reading that uncover the hidden power relations within imaginative works of literature – seems as over-schematic as the equal-and-opposite view that it is unconscionably naïve (or politically quietist) to look at patterns of linguistic diversity or complexity as these operate at the surface level of texts. Critique, I would also have thought, is as much needed as ever in the emergent (also emergency) conditions of the Anthropocene, which affects all of us but unevenly, and in which the social purpose of literary/cultural criticism – to pave imaginative ways to material transformation – is as relevant as it has ever been in the interconnected contexts of human-induced climate crisis, multiple species extinction, and disastrous environmental decline.

That said, several caveats are needed. The first and most obvious is that critique is not enough, perhaps especially those forms of cultural criticism that are associated with literary studies. All the literary criticism in the world will not stem the rising tides, and as Timothy Clark puts it in *Ecocriticism on the Edge* (2015), his landmark work on contemporary environmental criticism, there is a strong possibility that literary and other cultural critics may be deluding themselves into thinking that the "sphere of cultural representation is more important than it is" and, in amplifying the importance of the imaginary, are "running the risk of consolidating a kind of diversionary side-show" (21). A second caveat, linked to the first, is that the Anthropocene potentially complicates previous ways of reading, close and otherwise, to the point of invalidating them. As Clark suggests, the Anthropocene not only requires new ways of thinking our (human) relations to the planet, but also requires us to rethink the ways in which those relations are represented, insofar as Anthropocene concerns explicitly or implicitly "resist representation at the kinds of scale in which most thinking, culture, art and politics [currently] operate" – and have operated for some time (x). At worst, Clark implies, the Anthropocene is "unreadable," at least in the ways we normally read; at best, it requires a drastic readjustment of our reading practices in order to accommodate life – both human and nonhuman – at "much broader scales of space and time" (13).

Finally, a third caveat revolves around the question of the human. Insofar as literary studies is residually linked to humanist concerns, it is out of step with the kinds of radical recalibration (economic, ecological, epistemological) that the Anthropocene insists upon. As Rosi Braidotti argues, what is needed in the Anthropocene context is not simply a questioning of what it is to be human; it is a radical *undoing* of the human as a "normative category that indexes accesses to privileges and entitlements" (35). The Anthropocene, for Braidotti, is a "multilayered posthuman predicament that [encompasses] the environmental, socio-economic, affective and psychic dimensions of our ecologies of belonging," and this predicament requires nothing less than a new constellation of situated knowledges (Braidotti's umbrella term for this is "critical posthumanities") that work together without being assimilated to a universalist whole (32). Significantly, however, Braidotti does not part company with critique; on the contrary, she advocates for those kinds of "affirmative critique" (37) that bring together "radical epistemologies" (38), such as those associated with feminism, postcolonial studies, and critical race theory, which "take power seriously" and are firmly grounded in the present without losing sight of the past (39).

In what follows, I will largely side with Braidotti, although I am fully aware of the magnitude of the challenges posed by Clark. My more immediate aim is to compare two recent literary texts, both clearly written within the context of the Anthropocene, in which the interplay between "surface" and "deep" levels is literalized. This is not necessarily a defense of deep reading, which the two texts, albeit in different ways, implicitly pre-empt. However, as I hope to show, Robert Macfarlane's *Underland* and Will Hunt's *Underground*, both first published in 2019, are works that not only engage with the conditions of the Anthropocene, but also require forms of close reading that mesh with the radical epistemologies that

Braidotti references as a basis for confronting the Anthropocene predicament – epistemologies that reposition deep reading as a multilayered critique of modern technologies of extractivism rather than as a prescriptive form of extracting buried meaning from the text.

Underland

Underland is not Macfarlane's first encounter with the subterranean world. As he suggests in an op-ed piece in the *Guardian* published shortly before the official release of the book in May 2019, "our 'flat perspectives' feel increasingly inadequate to the deep worlds we now fashion and inhabit – and to the deep-time legacies we are leaving" ("Up from the Depths" 12). Hence his ongoing attempt to instill what the urban geographer Stephen Graham calls a sense of "verticality" in his writing, which is primarily evidenced in hard-core physical exploits such as climbing and caving, but is also apparent in his occasional forays into geology and, especially in *Underland*, his more sustained critical discussions of "the sunken networks of extraction, exploitation and disposal that support the surface world" ("Up from the Depths" 11; see also Graham). The eventual publication of *Underland* represents the culminating moment in what Macfarlane describes as a decade-long sequence of travels to places "where the underland deepens drastically or has exerted particular force on the upper world: from bronze age burial complexes in south-west England, to remote Arctic cave-art sites on Norway's northern coasts [the Lofoten Islands], and the blue depths of time archived by the ancient ice of Greenland" ("Up from the Depths" 10). Equally, as the book's subtitle, "A Deep Time Journey," makes clear, the book brings together its author's thoughts on deep time and its fraught relationship to the Anthropocene. Deep time, Macfarlane suggests, is nothing other than a "chronology of the underland . . . time measured in units that humble the human instant: epochs and aeons, instead of minutes and years" (*Underland* 15). The Anthropocene registers this time, but is perhaps better seen in terms of a radical scrambling of temporal coordinates: an uncanny presentiment of different timescales coming together or, still more, a disturbing sense of the untimely, of being in the wrong place at the wrong time, or of time itself running out (13–14). To put this in less saturnine terms, the Anthropocene demands of us that we think *ahead* in deep time and "weigh what we will leave behind as the landscapes we are making now sink into strata, becoming underlands" (77). "What is the history of things to come?" Macfarlane asks, and "what will be our future fossils? As we have amplified our ability to shape the world, so we become more responsible for the long afterlives of that shaping. The Anthropocene asks of us the question memorably posed by the immunologist Jonas Salk: 'Are we being good ancestors?'" (77).

In this context, *Underland* can be seen as a collapsing of deep (geological) time and long (human) history that corresponds to what new materialist theorists refer to as "storied matter": stories in which human and nonhuman bodies, and the supposedly separate worlds they inhabit, are mutually, indeed co-constitutively, entwined. Certainly, many of Macfarlane's more speculative contributions to the book could come straight out of the manual of new materialist theory, which insists on the liveliness of matter: "When viewed in deep time, things come alive that seemed inert. New responsibilities declare themselves. A conviviality of being leaps to mind and eye. The world becomes eerily various and vibrant again. Ice breathes. Rock has tides. Mountains ebb and flow. Stone pulses. We live on a restless earth" (*Underland* 16; see also Bennett, Coole and Frost).

By the same token, the philosophical aspects of Macfarlane's text need to be seen alongside, and in dialogue with, its physical dimensions, which include the potentially life-threatening challenges of caving in Somerset, crawling through the catacombs beneath Paris, and abseiling down a moulin (a meltwater-formed sinkhole) in Greenland, all of which offer opportunities for trademark exercises of partly self-disparaging Macfarlane bravado while confirming that he seems most at home in places that others would probably go out of their way to avoid. Close reading *Underland*, in this respect, does not seem to require much in the way of critical excavation: a surface reading, of the kind I have so far been performing here, is sufficient to give us access to its depths. However, Macfarlane also presents the underland in terms of what "cannot easily be seen or said: trauma, memory, grief, death, suffering, the afterlife" ("Up from the Depths" 10). What the underland hides from us may also be what we hide from ourselves; in this sense, it is frightening less because it is unknown to us than because, at some level, it *is* known to us: because of the secrets we have buried there that always risk eventually being unearthed. The underland, in this and other ways, is both a figurative place of death and darkness and an everyday terrain that lies literally beneath our feet; both a sanctuary where we "shelter what is precious (memories, precious matter, messages, fragile lives)" and a dumping ground where we "dispose of what if harmful (waste, trauma, poison, secrets)" (*Underland* 8). These dualities are maintained throughout the text, which repeatedly shows that the underland has both material and symbolic dimensions, and that its various sinister-looking portals – sunken caves, open tunnel-mouths, gaping moulins – offer precarious entry ways into a netherworld that holds real danger for us as well as haunting our imaginations in the shape of collective memories, shared metaphors, inherited myths (13).

Journeys to the world beneath, from Virgil to Verne, have long been a staple of the kind of philosophically inclined travel/nature writing at which Macfarlane excels, and it is to his credit that he is able to invest some of these well-worn subterranean narratives with new life.[1] One way he does this is to supply real-world examples that are almost guaranteed to be unknown to the majority of his readers: the appalling "death zones" of the Italian/Slovenian karst, for example, sites for a brutal twentieth-century clash between Fascists and Communists where sworn enemies and suspected collaborators alike were thrown in their thousands into secret limestone sinkholes (185); or the nuclear waste disposal site on Olkiluoto Island in south-west Finland, where a state-of-the-art bunker, recently constructed "to keep the future safe from the present" (403), comes equipped with warning signs in multiple languages, creating an uncanny atmosphere that resembles nothing less than an "Anthropocene black mass" (414).

Examples like this last one provide displaced reminders of Macfarlane's wide-angle critique of extractivism in *Underland*: those destructive processes, and the ideology that supports them, by which human beings are foolishly continuing to wager their futures against the future of a planet that will almost certainly outlast them, however many of its resources are taken from it, and however much it is technologically transformed. Again, deep reading seems not to be required on our part insofar as the text effectively does the job for us, with its "subsurface network of [intra- and intertextual] echoes, patterns and connections" (17) mimicking at the level of form what it also enacts at the level of content, and the text continually testing the boundaries of imagination in striving to find ever-new ways of describing itself. Crucially, *Underland* shows us what it purports to hide, potentially short-circuiting our investigative faculties as critics. Perhaps Moi is right: it is enough to be attentive to the text to unravel its mysteries, and it is the text itself rather than the fault-finding critic that dramatizes its own associations; that plays between its own constructed "surfaces" and its own equally fabricated "depths."

However, this *isn't* quite right any more than it would be right to accuse Macfarlane of empty formalism. There are still critical angles one can take to the text despite the fact that it often seems to be several moves ahead of us. A feminist approach is one of these. As I have discussed elsewhere, much of Macfarlane's writing is allied to the "extreme travel" genre, which is a by-product of what the German sociologist Ulrich Beck calls "world risk society" as well as an indirect expression of "residual masculinist anxieties at a time of sexual role redistribution and nuclear family decline" (*Extreme Pursuits* 6). The uncanny effects of Macfarlane's work are at least partly motivated by such gender anxiety, which surfaces from time to time in *Underland* in, for example, his ambivalent attitude toward the male subculture of urban exploration, which he at first admiringly defines as "adventurous trespass in the built environment" (153), only to skewer it later as a narcissistic form of "hipster entitlement" practiced by "obsessives . . . with tunnel vision" and given to sporting "Indiana Jones-style leather hats" (155, 170). Perhaps the most telling examples in the text, however, come in Macfarlane's interpolated descriptions of his relationship with his 4-year-old son, who is given pride of place in the book's poignant epilogue. Here, Macfarlane confirms his desire to be a good father as well as a "good ancestor," and the father's placing of his hand, palm to palm, against that of his son, which echoes a gesture previously made in the ancestral "Red Dancer" caves of the Lofotens, is the last of the text's uncanny encounters, physically rooted in place yet imaginatively traversing eons of deep time (425).

More could be said about why inheritance, which is key to the book's Anthropocene concerns, is framed here in male terms, and indeed about why Macfarlane's wife and, to a large extent, other women are conspicuously missing from his narrative world. As this critique has already been well made (Jamie; Bracke),

I don't intend to pursue it here, though it seems worth mentioning that Macfarlane's work rehearses a trope common to travel writing and nature writing alike: that of the absent or invisible (usually female) companion, whose support is vital to the whole enterprise but is not always readily acknowledged, and whose experiences are subordinated to what Timothy Morton calls the "masculinity memes" of the nature-writing genre ("Queer" 273). My point is hopefully made though: that a text that seemingly pre-empts critique also secretly invites it; and that a deep reading remains possible despite the text's attempts to bring "unreadable" Anthropocene tropes to the surface – to perform the unsettling effects caused by the shifting tectonics of space and the simultaneous extension and compression of (deep) time.

Underground

To read *Underland* and *Underground* in succession is to enter a strange kind of echo chamber in which the two texts begin uncannily to mirror one another: to resonate with each other's voices as well as with the numerous real/imaginary voices they incorporate, which stage – Sebald-style – a grand-scale colloquy between the living and the dead.[2] Both texts were originally published in the same year, and both represent *tour de force* physical and imaginative forays into underground spaces which double as unearthly "ghost landscapes," invisible to us on the surface but "unfolding everywhere beneath our feet" (5). Both cover a range of different examples, finding ingenious if sometimes questionable ways of making connections between them, and perhaps inevitably given their similar topic they sometimes turn to the same example, so that their recollections of, say, negotiating the catacombs of Paris become twice-told tales whose ramifying resemblances become simultaneously unnerving and involuntarily absurd. (Among the "Mole People" Hunt gets to meet in *Underground* is a homeless woman called Brooklyn, whose partner is also called Brooklyn, and who lives somewhere in the labyrinthine network of subway tunnels beneath Brooklyn, New York [22].)

Surface resemblances aside, *Underland* and *Underground* are very different kinds of text that adopt an equally different approach to their subject. *Underland*, as I have suggested above, is a philosophically inclined eco-travel narrative that engages more or less head on with the crisis conditions of the Anthropocene, asking what kind of world we have inherited from our forebears, but also what kind of world we will leave behind to those who follow us: like many an Anthropocene narrative, the tense it brings most readily to mind is the future past. *Underground*, however, appears little interested in either the future or the future anterior; instead, its main concern is with uncovering the archaic ancestral relationships to landscapes that Hunt sees as being wired into our nervous systems, relationships that connect us, across vast temporal and spatial distances, to our evolutionary past (24). Its main concern, in other words, is with human *origins*, and those origins are accounted for in a parallel series of creation stories and other mythological narratives that trace, not just the embeddedness of humanity within the larger earth system, but the emergence of humanity itself from underground (66, 84).

At its most basic level, *Underground* is an explicitly anthropocentric text, by no means uninterested in the behavior of other life forms but predominantly concerned with the underlying motives of *human* behavior, seeing – for example – the desire to dig as a "primal activity" by which we re-create our evolutionary connections to the earth (143). To tap into the earth, Hunt suggests, is to perform a kind of anthropological geology whose primary purpose is to lay bare the "elementary structures" of shared ancestral beliefs that remain "always intact in our core," however deeply buried these are (257). Hunt does recognize that tapping into the earth has other motives as well, for example, those revealed in what he sees as the sordid history of modern industrial mining (112), but this particular version of the Anthropocene story is buried in its turn beneath the *Ur*-narrative that drives the text, namely the ancestral search, embodied in secret/sacred underground practices and rituals, for "an older, earthier wisdom that runs deeper than logic and reason" – and to access this wisdom, "the route is down" (226). In fact, it is debatable whether *Underground* is an "Anthropocene text" at all and, unlike *Underland*, it certainly does not announce itself as such, though there are certainly Anthropocene *dimensions* to the text in its peeling back of modern history to reveal the cosmological grand narratives that precede it: narratives that are of a piece with Davies's view that the Anthropocene requires a radically altered understanding of the human relationship to *chronos*, new ways of "thinking with deep time" (12). In other respects, however, the text, for all its fascinating quests into the depths, offers a glittering play of surfaces, nowhere more so than in its tendency to see the narratives it assembles as revealing "innate biological rhythms" (205) that are common to *all* species, or as confirming practices that are integral to *all* civilizations and cultures, disclosing "the same essential ways of being in the world" (190).

A close reading of the text reveals surfaces rather than depths, counteracting its own literalized claims to be performing deep readings. This doesn't imply, though, that the text is somehow ineligible for a deep reading, or that a critical perspective on it is impossible in advance. To close this chapter, I want to pursue one particular line of critique, namely from critical race theory. Kathryn Yusoff's *A Billion Black Anthropocenes or None* (2018) brilliantly dissects the racial politics underlying the extractive industries and their discursive partner-in-crime, geology, which Yusoff sees as being complicit with a long history of racialized relations of power. For Yusoff, geology, ostensibly a neutral "language of rocks and deep time," uncovers an ongoing (colonial) history of racial exploitation, with the intertwined histories of extractivism (e.g., mining in the New World) and racial categorization (e.g., Blackness) both founded on the idea of inert/inhuman matter, to be acted upon and profited from by white men. Insofar as the Anthropocene invokes humanity as a geological force, Yusoff suggests, it erases the exploitative histories that make this "universalist geological commons" possible; what is therefore needed is less the kind of stratified reading (across time and space) that geology licenses than another kind of deep reading: of the fissions and fractures inherent in geology itself.

Yusoff's provocative thesis is far from failsafe, with perhaps its weakest point being its apparent (strategic?) need to assume one form or another of Anthropocene universalism. As Davies argues, "the Anthropocene does not require a turning away from the critique of sociopolitical power relations (globalization, capitalism, imperialism, and so on) toward a universal history of the human species. Instead, to understand the Anthropocene means widening the focus of sociopolitical critique and working toward an analysis of the power relations between geophysical actors, both human and nonhuman" (62). In other words, confronting the conditions of the Anthropocene involves a *joining together* of radical social critique and species thinking, which presents the opportunity to think about human and nonhuman relations at the same time (58). I am in agreement with Davies here, but this does not lessen the power of Yusoff's argument, which, in revealing the "Janus face" of geology – part "deep-time palaeontology," part "extractive economy" – resituates its "grammar of extraction" within the historical context of colonial theft.

This view can briefly be put to the test in a reading of one of Hunt's most arresting case studies, his short sojourn with the Wajarri people of Western Australia, which juxtaposes two mining histories – one modern and commercial (white-settler), the other ancient and alchemical (Aboriginal) – with the latter seconded to his main thesis that "for as long as we have extracted minerals from the earth, mining has been a spiritual act" (110). While Hunt's sympathies clearly lie with the Wajarri, his support is significantly weakened by his implication that modern industrial mining, rather than being exploitative in its own right, has lost touch with an ancestral earth history that encompasses all of humanity – then weakened further by the more hidden implication that the Wajarri, like other Aboriginal peoples, are threatened by a modernity on which they have effectively turned their backs. This last view not only flies in the face of the anthropological research that Hunt claims to be relying on, but also opens him to the criticism that (in Yusoff's words) he is willfully erasing a racialized history of mineral extraction that continues, all over the world, to "cut exposure along [geological] color lines."

There are significant limits to the particular genre that *Underground* is working within, the New Age spiritual travelogue, which tends to overlook cultural and historical differences in search of structural similarities, and which works from the premise that there is an atavistic (under)world that, largely unknown to us, reveals itself to us through "our ancestors' earliest dreams" (265). Hunt concedes, at the same time, that this world cannot fully be "read," and that "in our connection to subterranean space, we ease our suspicion of the unknown [while also recognizing] that not everything should be revealed, not all the time" (265). The underground, Hunt concludes, "teaches us to respect mystery. . . . [It] reminds us of what our ancestors always knew, that there is forever power and beauty in the unspoken and the unseen" (265–66). Sententiousness aside, pronouncements like this are not entirely removed from the "uncanniness" of the Anthropocene, which registers the limits of human knowledge, and which repositions us within a vast "network of evolutionary developments and ecological interactions," over only some of which we (whoever "we" are at any given moment) can stake a legitimate claim to having influence or control (Davies 76). Nor is Hunt's work, oriented toward a popular readership though it is, necessarily out of step with more idiosyncratic approaches, notably Donna Haraway's, which mobilize ancient chthonic powers against an Anthropocene that seeks to exterminate them: "chthonic ones," in Haraway's parlance, that "writhe and luxuriate in manifold forms and manifold names in all the airs, waters, and places of earth" (2).

The Anthropocene *is* mysterious in some ways, though that doesn't mean that some of the discourse surrounding it can't or shouldn't be demystified, still less that the different knowledges and knowledge formations that have been applied to it all belong to the same base. To return to Braidotti, the Anthropocene requires forms of "affirmative critique" that disidentify from the potentially all-encompassing tropes of Man and Anthropos, and that bring together the "radical epistemologies" that are needed to confront a crisis which casts doubt, in turn, on some of the categories that generations of literary/cultural scholars have rallied around: categories of "nature," categories of "culture," categories of the "human" itself (37–38). And if the human has been brought to crisis by the Anthropocene, so too has the humanist apparatus of close reading, which Love sees as being embedded in our "institutional DNA" as our critical mission to produce readings that attribute "life, richness, warmth, and voice to [literary] texts" (388). Can new kinds of close reading emerge that are able to shake off these humanist legacies? My own view is that they can and have, but not in the form of "surface" reading, rather in those judicious forms of "deep" Anthropocene reading that are rightly suspicious of anthropocentrism while remaining committed to the social purposes and political imperatives of critique.

Notes

1 There is no space here to discuss whether Macfarlane is better seen as a travel writer or, as he himself prefers, a nature writer: most of his writing to date contains elements of both. Macfarlane has also associated himself with the so-called new nature writing in the United Kingdom, a particular brand of environmentally conscious writing that combines critical self-reflection, philosophical speculation, and lyrical descriptions of the natural world. For introductory insights into the new nature writing, see the essays in Cowley; for a more extended discussion, see also Smith. As Macfarlane among others has been quick to recognize, the Anthropocene poses a massive challenge to nature writing, and the ecocriticism that seeks to account for it, insofar as the "nature" that previous writers have drawn upon is no longer recognizable; indeed, the very term "nature" might be considered potentially destructive as well as alarmingly out of date. For alternative views on this seemingly intractable dilemma, see Clark ("Climate"), Morton (*Ecology*), and Purdy.

2 Hunt's formal indebtedness to Sebald is not difficult to see, from his dense network of (trans)cultural associations to the grainy black-and-white photographs that are interspersed throughout his work. The content of *Underground*, too, is partly Sebaldian in its

contemplation of the "invisible forces that shape our reality" (30), though Hunt's work lacks the mournfulness of Sebald's, coming closer in its engaging tone and jackdaw philosophizing to the travel narratives of Bruce Chatwin, with whom Hunt also shares a fairly cavalier approach to his source materials as well as his presentation of cultural belief systems and historical facts.

Works cited

Anker, Elizabeth S. and Rita Felski. "Introduction." *Critique and Postcritique*. Ed. E.S. Anker and R. Felski. Durham, NC: Duke University Press, 2017. 1–28.

Beck, Ulrich. *Risk Society: Towards a New Modernity*. London: Sage, 1992.

Bennett, Jane, *Vibrant Matter: A Political Ecology of Things*. Durham, NC: Duke University Press, 2010.

Bracke, Astrid. "Macho Nature? Or, Gender in New Nature Writing, Part 1." Web. 25 April 2019. https://astridbracke.com/2013/02/19/macho-nature-or-gender-in-new-nature-writing-part-I.

———. "Typically Feminine? Or, Gender in New Nature Writing, Part 2." Web. 24April2019.https://astridbracke.com/2014/01/02/typically-feminine-or-gender-in-new-nature-writing-part-2.

Braidotti, Rosi. "A Theoretical Framework for the Critical Posthumanities." *Theory, Culture & Society* Vol. 36.6 (2019): 31–61.

Clark, Timothy. *Ecocriticism on the Edge: The Anthropocene as a Threshold Concept*. London: Bloomsbury, 2015.

———. "Some Climate Change Ironies: Deconstruction, Environmental Politics, and the Closure of Ecocriticism." *Oxford Literary Review* Vol. 32.1 (2010): 131–49.

Coole, Diana and Samantha Frost, eds. *New Materialisms: Ontology, Agency, and Politics*. Durham, NC: Duke University Press, 2010.

Cowley, Jason, ed. *The New Nature Writing*. London: Granta, 2003.

Davies, Jeremy. *The Birth of the Anthropocene*. Oakland, CA: University of California Press, 2016.

Felski, Rita. *The Limits of Critique*. Chicago: University of Chicago Press, 2016.

Graham, Stephen. *Vertical: The City from Satellites to Bunkers*. London: Verso, 2016.

Haraway, Donna. *Staying with the Trouble: Making Kin in the Cthulucene*. Durham, NC: Duke University Press, 2016.

Harvey, David. *The Condition of Postmodernity: An Enquiry into the Origins of Cultural Change*. Oxford: Wiley-Blackwell, 1992.

Huggan, Graham. *Extreme Pursuits: Travel Writing in an Age of Globalization*. Ann Arbor, MI: University of Michigan Press, 2009.

Hunt, Will. *Underground: A Human History of the Worlds Beneath Our Feet*. New York: Random House, 2020.

Jamie, Kathleen. "A Lone Enraptured Male." *London Review of Books* Vol. 30.5 (2008): 25–27.

Latour, Bruno. "Why Has Critique Run Out of Steam? From Matters of Fact to Matters of Concern." *Critical Inquiry* Vol. 30 (2004): 225–48.

Love, Heather. "Close but not Deep: Literary Ethics and the Descriptive Turn." *New Literary History* Vol. 41.2 (2010): 371–91.

Macfarlane, Robert. *Underland: A Deep Time Journey*. London: Hamish Hamilton, 2019.

———. "Up from the Depths." *Guardian Review* [London]. 20 April 2019: 6–13.

Miller, D.A. *Jane Austen, or the Secret of Style*. Princeton, NJ: Princeton University Press, 2003.

Moi, Toril. "'Nothing is Hidden': From Confusion to Clarity; or; Wittgenstein on Critique." *Critique and Postcritique*. Ed. E.S. Anker and R. Felski. Durham, NC: Duke University Press, 2017. 31–49.

Morton, Timothy. *Ecology without Nature: Rethinking Environmental Aesthetics*. Cambridge, MA: Harvard University Press, 2009.

———. "Queer Ecology." *PMLA* Vol. 125.2 (2010): 273–82.

Purdy, Jedediah. *After Nature: A Politics for the Anthropocene*. Cambridge, MA: Harvard University Press, 2015.

Rooney, Ellen. "Symptomatic Reading is a Problem of Form." *Critique and Postcritique*. Ed. E.S. Anker and R. Felski. Durham, NC: Duke University Press, 2017. 127–52.

Smith, Jos. *The New Nature Writing: Rethinking the Literature of Place*. London: Bloomsbury, 2018.

Yusoff, Kathryn. *A Billion Black Anthropocenes or None*. Minneapolis: University of Minnesota Press, 2018. Web. 7 October 2020. https://manifold.umn.edu/projects/a-billion-black-anthropocenes-or-none.

10

READING IN THE DARK

The aura of eclipse

Hilary Thompson

Imagine somewhere in the United States a holiday dinner table where the political positions of family members seated around it run from starkly partisan, agnostically populist, to variously libertarian. It becomes impossible to discuss particular candidates for an upcoming presidential election with any kind of agreed upon knowledge or interest because people have few media outlets in common. But it comes out that nobody likes the Electoral College.

This could be an analogy for the current situation in academic literary scholarship in which, not often reading the same texts or having the same reasons for reading the texts that we do, we turn to method as the topic in common. Within this topic, forms of reading, seemingly our most basic activity, command a special appeal. And within this area, one can almost measure trends by their attraction to or repulsion from those magical words, "close reading." As with "critical thinking," it's difficult to gainsay the value of terms suggesting sharp cognition and focused attention. But in the arc of a semester, we may move toward greater awareness of contextual and not specifically text-based phenomena, thus incorporating steps toward the disenchantment of the term whose magic we initially needed. While in our official duties it may be as hard for us to fully abjure close reading as it is for us to dispense with reading, when we turn to writing, so many new terms can efface "close" to preface "reading." It's as though we want new spells in which "pocus" or "corpus" could come after anything other than "hocus" or "hoc est."

Literary readings of heavenly bodies, particularly the dramatic instance in which a specific alignment briefly swallows us in darkness, reveal the paradoxes built into our ambivalent desires for both closeness and distance in interpretation. The idea of Anthropocene and the total solar eclipse both offer a glimpse of planetary life's extinction; in the context of the former, the latter becomes more significant, as the basis for spectacular creativity, for a concentration of meaning in which the very terms of literariness come to light. This essay unfolds in three phases as it makes the case that eclipse writings offer a unique opportunity to practice a form of conjunctive reading in which surfaces overlap and the near and far may be meaningfully contemplated at once. First, I consider recent revisitings of close reading that offer potential for its historical and cosmic enrichment; second, I make a brief foray into forms of nature writing, from forest biology and cultural anthropology, that are informed by and transform notions of close reading; and third, I read modern literary treatments of the eclipse experience in which close observation, forms of unseeing, and serendipitous association converge to suggest alternative logics, ones in which the disjunctive leap and forces beyond the human predominate.

Aura: closing the distance or toward a truly close reading

In "The Work of Art in the Age of Its Technological Reproducibility," Walter Benjamin famously defined the aura of original artworks (as opposed to their aura-lacking reproductions) as a "strange tissue of space and time: the unique apparition of a distance, however near it may be" (104–5). The same paradoxical quality, Sean Michael DiLeonardi (2019) leads us to conclude, could be ascribed to reading. It's not unusual to hear scholars gently mock the movement historically associated with close reading, the New Criticism. We remember school essays we once wrote, ones that claimed the recurrence of this or that image, such and such a word in a short story or poem couldn't simply be "by accident" or "just a coincidence." Or we think of essays students may write to please us that claim readers need to "look deeper" to discover the "hidden meaning." DiLeonardi usefully outlines the history by which the New Criticism's focus on textual interpretation as the recognition of patterns so statistically significant as to be impossible to attribute to mere coincidence emerged in tandem with and influenced by mid-century advances in code cracking and the attempt to apply similar principles to translation – both with recourse to mathematics and the potential help of machines. Arguing that in the post-WWII years "mathematic models of language and traditional close reading both developed through applications of cryptography to language," which itself relied on quantitative calculations to differentiate significant patterns from coincidental repetitions, DiLeonardi unearths a history that disciplinary boundaries and digital phobias have hidden. He notes that William F. Friedman, the famous cryptologist and author of *The Index of Coincidence and Its Applications in Cryptanalysis*, not only dabbled in literary criticism but was consulted by none other than W. K. Wimsatt as he wrote the core New Critical text, "The Intentional Fallacy." DiLeonardi focuses particularly on that essay's key inheritance from cryptography: its dependence on a pattern/coincidence distinction and its use, whether interpreting one text or comparing two or more, of "a method [that] relies on the consistencies provided by close readings of the actual texts." In essence, we have an empirical, object-based, quantitative method, one amenable to a digital approach. DiLeonardi comes to the conclusion that despite the divergent rhetoric between computational cryptography-friendly theorists such as Wimsatt and such stalwart humanists and advocates of literary difficulty as Cleanth Brooks, "Close and distant reading, not so separate after all."

"Close reading" names a set of paradoxes, as though in practicing it, interpreters and their texts were never really that close but were, unbeknownst to even themselves, in a distant relationship. In Benjamin's terms, we might seem to have a reverse aura. It is historically ironic that the literary interpretive method most invested in the text's very words, in whatever edition we happen to encounter them, and in this sense, most invested in the text's aura, should have come to be defined also by its investment not in distance but in closeness (the supposed closeness of the text however distant its design may seem). Recent methods such as Franco Moretti's, which return the analytical gaze to the distance, to macro patterns of large datasets, do little to change the basic nature of this aura, as they merely return close reading to its distant roots. As he says, no less paradoxically, of the difference between canon-driven close reading and his proposed "distant reading" of world literature:

> At bottom it's [close reading] a theological exercise – very solemn treatment of very few texts taken very seriously – whereas what we really need is a little pact with the devil: we know how to read texts, now let's learn how *not* to read them. Distant reading: where distance, let me repeat it, *is a condition of knowledge*: it allows you to focus on units that are much smaller or much larger than the text: devices, themes, tropes – or genres and systems. And if, between the very small and the very large, the text itself disappears, well, it is one of those cases when one can justifiably say, Less is more.
>
> *(57)*

In disenchanting select texts of their distinctiveness to instead track systems of variation across numerous examples of world literature through various eras, his distant reading takes a form of textual unseeing as its apparent premise, yet it is the act of meaningfully grasping counterintuitive patterns that brings the point of Moretti's method home. If what we call "close reading" was similarly bolstered by a belief in a not immediately accessible design, yet one that might be gleaned from forms of data-driven analysis, then we can conclude that its sense of textual presence indeed was, as with Benjamin's original artwork and its aura, a quality of distance, an original distance that close reading's appeal to closeness merely accentuated. If distant reading is close and close reading is distant, we are left with the question of what truly close forms of reading would look like, that is, interpretive techniques premised on the meaningfulness of nearness itself.

Simone Stirner has recently invoked Benjamin to propose "a technique of closeness." Following Sharon Marcus and Stephen Best's "surface reading" (reading that eschews the search for deeply hidden meanings and contrary-to-appearances subtexts), Stirner draws on Benjamin's discussions of flânerie and embodied perception to outline a theory of "closeness reading." While Stirner sees "surface reading" as an "umbrella term" for various approaches, she sympathizes with its general impulse to "return to the surface of a text, be it in the form of a text's materiality, its affective and sensuous dimensions, or anything else that is taken to be 'there' instead of hidden and repressed" (273). Seeking to enhance surface reading techniques with greater potential for historical depth, Stirner offers a flâneur-inspired style of sympathetic surrender to associative vagaries, claiming, "Flânerie – in its rhythmic walk and free-floating attention that does not decode the city but enters into an intimate nearness with it – appears as a surface reading of a different kind, one in which the contact with the surface of a city might suddenly open up to deep time, emergent configurations, and cross-historical formations" (277). For Stirner, Benjamin's opening description of the flâneur in Convolute M of *The Arcades Project* beautifully encapsulates this approach:

> But the great reminiscences, the historical shudder – these are a trumpery which he (the flâneur) leaves to tourists, who think thereby to gain access to the genius loci with a military password. Our friend may well keep silent. At the approach of his footsteps, the place has roused; speechlessly, mindlessly, its mere intimate nearness gives him hints and instructions. He stands before Notre Dame de Lorette, and his soles remember: here is the spot where in former times the *cheval de renfort* – the spare horse – was harnessed to the omnibus that climbed the Rue des Martyrs toward Montmartre. Often, he would have given all he knows about the domicile of Balzac or of Gavarni, about the site of a surprise attack or even of a barricade, to be able to catch the scent of a threshold or to recognize a paving stone by touch, like any watchdog.
>
> (Arcades *416)*

Analyzing this passage, Stirner notes a fundamental duality between the "trumpery" by which tourists access grand history with "a military password" and the flâneur who senses past-lived details from the bottom of his feet. It is the "mere intimate nearness" of this sensuous, embodied apprehension of the past amidst the present, this conjoining of times, that we can see in the phrase "*here* is the spot where" (emphasis mine), that inspires Stirner's Benjaminian notion of closeness reading. But also important here, especially for modes of reading seeking to move beyond anthropocentrism toward multispecies perspectives, is Benjamin's appeal to the omnibus's spare horse and the acute sense of the watchdog. Benjamin charts a move beyond code and coded history toward ethology and ecology.

Stirner's Benjaminian approach can be further expanded because the "deep time" and "emergent configurations," or in Benjamin's favored word, "constellations," it opens up immediately suggest potentially cosmic dimensions. If we see distant reading and close reading in our time as not entirely good faith versions of one another, then, following Benjamin and Stirner, a mode of reading that takes the experience of near surfaces as the basis for far-off associations, and does so by design, might become not only esthetically and ethically preferable, but particularly useful in the era of the Anthropocene, that epoch whose perception turns on our seeing our material lives in vast geological and deep temporal terms. Taking a cue from Stirner, we can bring Benjamin's suggestions toward a truly close reading practice, one with potentially cosmic reach, into constellation with essays describing the distant, yet close event of a total solar eclipse. While the heavenly bodies brought into alignment in this event are astronomically far from us and each other, it's from our vantage point on the Earth that we perceive their aligning and its effect of shadowing. Whether we read the event as a catastrophic effacing or primal conjoining depends on our ways of interpreting our cosmic position. Indeed, what comes together in writings of the eclipse is a ground-up tradition of observation-based nature writing, one of particular importance for considerations of the Anthropocene, and a surreally speculative technique of montage. On the one hand, we have a cultivation of expanded attention that paradoxically relies on forms of unseeing. On the other, we have an intense interest in charged connections, ones that seem outside the bounds of logic. A practice of conjoint reading allows us to view these strands together, with an eye for the ways that the total solar eclipse suggests a glimpse of our destruction but also a chance to learn ways of seeing ourselves in darkness.

Earth: reading from the ground up

As the humanities have renewed their relationship with the computational, branches of the sciences and social sciences, especially those concerned with ecology, have reached over into the literary. As we'll see, these scientific literary forays bear a familial resemblance to canonical nature writing, in particular the work of Annie Dillard. Both forest biologist Robin Wall Kimmerer and cultural anthropologist Anna Lowenhaupt Tsing have not only written beautifully poetic prose on such earthly phenomena as species of moss and a special variety of mushroom (the prized matsutake) but also they have advocated a method that looks startlingly like literary close reading. In the brief personal essays of *Gathering Moss*, Kimmerer mixes elements from her knowledge of botany and her indigenous American heritage to describe mosses' complex world. Perceiving mosses as she enjoins us to requires an unusual perceptual capacity, and she educates readers in it with a chapter called "Learning to See." Likewise Tsing entitles her first chapter in *The Mushroom at the End of the World: On the Possibility of Life in Capitalist Ruins* "Arts of Noticing." Both their approaches resonate with Stirner's nearly tactile sense of closeness reading, with Benjamin's calls for heightened, anti-anthropocentric attention and Dillard's highly literary ways of reading natural environments.

Kimmerer's approach might initially seem at odds with the flâneur's, as she admits, "I go to the city whenever I must and leave as soon as I can" (98), yet, as Benjamin does, she finds productive value in strolling and urban observations.

She notes that city environments support the silvery moss *Bryum argentum* through "dispersal by footsteps" because its "shoots are fragile at their tips and in fact designed to break off," which means, "The broken tips, scuffed along by a pedestrian, will take hold in another sidewalk, spreading Silvery Bryum all over the city" (93). If the flâneur remembers with the soles of his feet, the oblivious pedestrian's help disperse and propagate, rerouting and rerooting at once. Likewise, although quickly glossing over the different mosses on a forest log is certainly a crime for Kimmerer, it's one of poor flânerie that she puts in terms any cultured urbanite can understand: "To pass hurriedly by without looking is like walking by the Mona Lisa chatting on a cellphone, oblivious" (10). By contrast, her mode of learning to see is an enhanced flânerie that entails acknowledging and overcoming one's distance, one's inappropriate sense of scale, but also surrendering to straying vision:

> At the scale of moss, walking through the woods as a six-foot human is a lot like flying over the continent at 32,000 feet. So far above the ground, and on our way to somewhere else, we run the risk of missing an entire realm which lies at our feet. Every day we pass over them without seeing. Mosses and other small beings issue an invitation to dwell for a time right at the limits of ordinary perception. All it requires of us is attentiveness. Look in a certain way and a whole new world can be revealed.
>
> *(10)*

The path to this "certain way," however, may not be direct. Recalling her first attempt to see starfish in tidepools that was abundantly successful only when she stopped looking and stood up to relieve back pain, she realizes she had to *learn* to see them in a new environment's shapes and colors. She then cites a bit of counterintuitive wisdom she was once taught:

> A Cheyenne elder of my acquaintance once told me that the best way to find something is not to go looking for it. This is a hard concept for a scientist. But he said to watch out of the corner of your eye, open to possibility, and what you seek will be revealed. The revelation of suddenly seeing what I was blind to only moments before is a sublime experience for me. I can revisit those moments and still feel the surge of expansion. The boundaries between my world and the world of another being get pushed back with sudden clarity, an experience both humbling and joyful.
>
> *(9)*

The sublime arrives serendipitously as a revelation amidst blindness as one lets go of habitual perceptions and perceptual expectations. Learning to see is first learning to unsee.

This same notion of enhanced flânerie in which the wandering yet attentive stroll serves as model for broader worldly awareness informs Tsing's arts of noticing. She opens her book's prologue with the rhetorical question and answer: "What do you do when your world starts to fall apart? I go for a walk, and if I'm really lucky, I find mushrooms. Mushrooms pull me back into my senses, not just – like flowers – through their riotous colors and smells but because they pop up unexpectedly, reminding me of the good fortune of just happening to be there" (1). Tsing doesn't set out looking for mushrooms; they are just "there," fortuitous finds. The sense with both Kimmerer's cellphone and Tsing's flowers is that when instrumentalist perception is set aside, mosses and mushrooms emerge and emblematize new networks. They are models of the "'search image' in the brain" we've just learned to recognize (Kimmerer 9); the food you can forage for when "the controlled world we thought we had fails" (Tsing 2); and the wild trade item that, along with the surge in diverse foraging traders entering forests, goes unremarked as people who discuss deforestation rehash the economy versus environment debate (Tsing 18). Tsing's prized yet wild and uncultivatable mushrooms, specifically the matsutake, which grow in damaged landscapes, even nuclear blast sites, are like Kimmerer's mosses, anomalous objects that point up what we habitually don't and now need to see. For both writers, the networks of life emerging from shadowy surfaces, whether cracked pavement, fallen logs, tidepools, or forest soil, are vital models of systems whose significance calls for close, but reconfigured attention. As Tsing says, "we must revitalize arts of noticing," moving across traditional disciplinary restrictions when the "research object is contaminated diversity," and "its unit of analysis is the indeterminate encounter" (37). Such indeterminate encounters occur not only to roaming flâneurs, but to those who let their eyes wander far overhead, scanning the surface not of the pavement but of the firmament.

If both Kimmerer's and Tsing's forms of earthly attention resemble reading, it is likely because they inherit a mode of nature writing, a spiritually-inflected one, from Dillard, who draws on the transcendentalists and religious tradition. This is most obvious in the way that Kimmerer's "Learning to See" echoes Dillard's "Seeing" in her 1975 Pulitzer prize-winning book *Pilgrim at Tinker Creek*. Here she explores the rich life around this creek, lingering there even in darkness. Meditating on the strange forms of sight her excursions foster, she recalls once searching for a particular tree, and finding it (predictably for us) only when she wasn't looking. This anecdote unfolds on the heels of her consideration of those who have had their blindness from cataracts surgically cured: "Many refused to use this incomprehensible new sight of theirs. Others, who had known objects only by feel, began *learning to see*" (171, emphasis mine). Where Kimmerer cites a Cheyenne elder for perceptual guidance, Dillard quotes turn-of-the century spiritualist and nature writer Stewart Edward White, who urged would-be deer viewers, "As soon as you can forget the naturally obvious and construct an artificial obvious, then you too will see deer" (qtd. in Dillard 156). Seeing becomes almost providential in Dillard, an act of faith, and her observation-based nature writing becomes a crucial precursor for Kimmerer's and Tsing's cultivated arts of seeing in which vision is undone and then retrained, veered away from well-worn instrumentalist tracks toward forms of close, almost tactile apprehension or else distant reconfigurations. It's as though forms not just of literacy but of high literateness inherited from some of the most theological nature writers are reverse engineered to take in expanded apprehensions of what lies beyond or at the edges of habitual human-centric vision.

Eclipse: toward conjunctive reading

Strangely crucial to this method, in which closely akin writers poetically plead for enhanced perception, is a technique of non-citation in which authors retrace precursors' outlines without always saying so. Just as Moretti notes of distant reading's tracking of variants, "between the very large and the very small, the text itself," in this case of significant forebears, "disappears." While one might not expect self-conscious and highly literary intertextual gestures from writers in far-flung fields, Dillard, too, fits into this tradition of silent homage despite her writing's many explicit literary references. Her chapter "Seeing" launches the technique she later uses in her much anthologized 1982 essay "Total Eclipse," to the point that for her an eclipse viewing becomes the ultimate instance of retrained vision: "Usually it is a bit of a trick to keep your knowledge from blinding you. But during an eclipse it is easy. What you see is much more convincing than any wild-eyed theory you may know" ("Total" 8). But as Jocelyn Bartkevicius demonstrates, "Total Eclipse" also silently retraces the outlines of another eclipse essay, Virginia Woolf's 1928 "The Sun and the Fish." If the earthly close reading of nature writing generally thrives on close networks it only partly brings to light, the tradition of eclipse writing presents an especially heightened example of this pattern.

Bartkevicius describes discovering this Dillard-Woolf overlap when, teaching a course on nature writing, she left the confines of familiar expository essay anthologies, ones that included Dillard's "Total Eclipse," to find writing that would show the mind at work, with all its disjunctive leaps and associations. She arrives at a text we could see as a prime example of high modernist literary difficulty and hence a prime candidate for New Critical close reading, Woolf's "The Sun and the Fish." Bartkevicius sees immediately that it's also an eclipse essay, but then, along with her class, discovers it establishes a point-by-point pattern for Dillard's "Total Eclipse" to follow, a pattern with points too numerous to seem coincidental. Among the many shared features she and her students observe, she highlights several parallel retrospective descriptions: of the journey to the eclipse viewing, of the strange crowds drawn together and their erased individuality, of the initially cloudy sky, of the suddenly blacked out sun and fast-moving line of shadow, of the people seemingly rendered inanimate, of the intimation of apocalypse, and of the thoughts of ancient sites (45–46). Yet despite their overwhelming number of points in common, the essays, for Bartkevicius, have radically different methods and ends. Dillard's remains a piece of orderly expository writing, moving chronologically, laying out all its key symbols early so as to flesh out their larger meanings in due course, and culminating in a very human epiphany in which "what the narrator apparently mourns most of all is the possible loss not of the entire earth, but of human life and community in particular." Woolf, by contrast, is not wed to chronology or human significance. Rather, Bartkevicius argues, she is interested in forces beyond conscious human control, whether involuntary mental associations or planetary movements:

> It is some mysterious, complicated, and hidden aspect of mind that associates images, that puts together objects in the mind and memory of a person. This aspect of mind not only resembles the natural world but is affected by it as well. The changes surrounding the eclipse awaken a new way of seeing – the motion of the sun affects the motion of the mind.
>
> *(47)*

Woolf succeeds in modeling another form of writing for Bartkevicius, one that opens up the possibility of greater engagement with the nonhuman, because she channels the power of disjunctive association.

Likewise, mushroom forms and eclipse images become accidentally associatively linked in the work of both Tsing and Dillard. While the matsutake becomes the emblem of Tsing's book on collaborative survival amidst the capitalist ruins of the Anthropocene, her follow-up coedited volume *The Arts of Living on a Damaged Planet* has part of its origin in an eponymous conference whose website bore another emblem: Liz Millar's photo of a total solar eclipse.

Visually stunning, the conjunction of an eclipsed sun with the concept of the Anthropocene inspires questions. While one could imagine, for example, a symposium on nuclear proliferation making use of posters showing a mushroom

FIGURE 10.1 *The Arts of Living on a Damaged Planet* conference website.

cloud, this conference's use of an event not caused by humankind and not occurring from all earthly viewpoints is intriguing. By what logic do we take the natural event of such a celestial alignment, viewed from one perspective, as evocative of life on our damaged planet? Distantly caused yet almost tangibly darkening, the total solar eclipse represents a concentrated convergence of the forms of significance pursued by various methods of reading: like the New Critical text, it suggests paradoxes; like distant reading's dataset, it makes visible hidden anomalies; like surface reading's undeniably "there" significance, it insists on its presence; like closeness reading's sense of the very spot "here" as the where of long past phenomena, it opens up portals to the far-off; and like earthly close readings' scrutiny of shadowy surfaces, it unearths new routes of perception. If we take our own perspective for granted, we view it not as a passing shadow, but a dark entity that reflects our faults as nations and as a species. If we let ourselves be thrown off-center, we learn to see new connections, ourselves as members of an unaccustomed commons.

We can see this tension between re- and decentering at play not only, as Bartkevicius does, in eclipse essayists' most apocalyptic moments, but also in their most eccentric connections. And, as with eclipses themselves, what seems most centrally significant from one perspective might seem most peculiar, and historically so, from another. Analyzing at greater length the moment that Bartkevicius sees as Dillard's key human-centered epiphany as she views the 1979 eclipse, we might note also its Cold War ethos:

> Seeing this black body was like seeing a mushroom cloud. The meaning of the sight overwhelmed its fascination. It obliterated meaning itself. If you were to glance out one day and see a row of mushroom clouds rising on the horizon, you would know at once that what you were seeing, remarkable as it was, was intrinsically not worth remarking. No use running to tell anyone. Significant as this dread sight was, it would not matter a whit. For what is significance? It is significance for people. No people; no significance. This is all I have to tell you.
>
> *("Total" 13)*

Describing an eclipse that occurred in the same year as the release of *Apocalypse Now*, Dillard's essay is riddled with imagery suggesting war and nuclear holocaust, as though America's superpower identity looms large. Before mushroom clouds come to mind, she refers to the crowd as "look[ing] as though we had all gathered on hilltops to pray for the world on its last day. It looked as though we had all crawled out of spaceships and were preparing to assault the valley below" (6). She depicts her then husband Gary in the eerie eclipse light, noting "The skin on his face moved like bronze plating that would peel" (10). And she remembers the whole sight as though a vision from the other side of life, but likens the death vision to one a wartime annihilation would cause: "What I saw, what I seemed to be standing in, was all the wrecked light that the memories of the dead could shed upon the living world. We had all died in our boots on the hilltops of Yakima, and were alone in eternity" (16). Dillard offers cautionary words when she compares innocuous photographs of eclipses to Christmas cards and opposes them to the sight of the real thing: "More moving photographs than those of the sun's corona can appear in magazines. But I pray you will never see anything more awful in the sky" (14). Not only does "Total Eclipse" center itself within the realm of human significance but also the reader for whom it strains to make an eclipse experience real is a distinctly American one. The awe in which the essay holds this astronomical phenomenon is implicitly likened and linked to the awe Americans should feel at their nation's powers of destruction.

While the total solar eclipse and the mushroom cloud (along with the mushrooms that can emerge after it) might substitute for each other in an apocalyptic logic that moves from the Cold War to ecological crisis, the eclipse's iconic significance is not exhausted by this progression. Although such iconic sights of the extinguishing of life and light leave Dillard with "nothing more to tell," for Bartkevicius, Woolf indeed "has more to tell us" because for her "the earth's significance . . . does not depend on the presence of people" (47). And this is seen in Woolf's abrupt jump from contemplating human extinction to the then new aquarium of London's Zoological Gardens and her extended comparison of living human form with that of lizards and fishes. Curiously, a further Dillard–Woolf similarity, one not noted by Bartkevicius, is Dillard's description of an aquarium: "On the broad [hotel] lobby desk, lighted and bubbling, was a ten-gallon aquarium containing one large fish; the fish tilted up and down in its water. Against the long opposite wall sang a live canary in its cage" ("Total" 4). Dillard clearly conjures a zoo-like atmosphere, but these animal images occur before the real expedition begins, and even observing the birdseed spilled on the floor, she seems more interested in contemplating animal containment than animality at length. For Woolf, apocalyptic imagery appears en route to assessing the human form from the reference point of nonhuman animal bodies, whereas for Dillard the aquarium and canary cage remain eccentric images popping up on the way to the hilltop viewing and restaurant breakfast after. While she expands her expository essay's aquatic imagery to contemplate highly metaphorical deep-sea dives for pearls and treasure amid mysterious beasts, the contemplation of the nonhuman itself is not the goal.

Yet it is precisely Woolf's extended meditation on the nonhuman that puzzles poet Anne Carson in her essayette "Totality: The Colour of Eclipse," first published in 2003. Here she returns to the question of conjunction itself, as she looks back on multiple writings of the eclipse. Carson leaves no reasonable forebear uncited, her four and a half pages addressing ancient Greek word roots, Herodotos, Germanic legend, Emily Dickinson, and Dillard, but primarily Woolf. In the midst of these textual readings, distant and close, what has disappeared is firsthand observation of the phenomenon itself. Carson's reading of eclipse is a reading of reading, all its significant connections in plain sight, but the central event of darkening left out. What draws her interest is the possibility, found in eclipse descriptions, that a seeming void can have a color, lack of substance a tangible quality, and a world thrown into its inverse state, an unusual logic of connection and desire. Just as Woolf's end image is Dillard's eccentric one and it is Dillard's apocalyptically military moments that might now read as eccentric, where one interpreter could read eclipse writings for new imaginings of extinction and ecology, that is, signs of awareness of the Anthropocene, Carson reads for queerness, for the non-heterocentric.

For Carson, the consummately odd image in Woolf's essay is the sun's partner in her title, the fish. Carson's essayette, somewhat whimsically and willfully, begins with the claim, "You might think a total solar eclipse would have no colour," but then counters it with the assertion, "Yet people who experience total eclipse are moved to such strong descriptions of its vacancy and void that this begins to take on colour" (149). This notion, more accurate to Dillard, surely comes partly from Woolf's epiphanic contemplation of extinction (that Dillard's death vision follows and modifies), but there Woolf seems to say the opposite of Carson's assertion: "But still the memory endured [after the sunlight's return] that the earth we stand on is made of colour; colour can be blown out; and then we stand on a dead leaf; and we who tread the earth securely now have seen it dead" ("Sun" 90). But core to Carson's piece is not so much close reading as the value of the anomalous that she finds scanning a large dataset to support. She revels in the concept of reversal and value of revelations occurring amid a seemingly wrong world because "Wrongness has its own colour and it is not like anything else" (150). Cross-referencing "The Sun and the Fish" with Woolf's diary entry detailing her eclipse expedition with husband Leonard, lover Vita Sackville-West, and Sackville-West's husband Harold Nicholson, Carson has an eye for oblique signs of desire. So when Woolf's essay moves from the vision of a dead earth to the London Zoo seen in the mind's eye, Woolf's first sight there – copulating lizards – comes as no surprise. Carson accepts the essay's explanation that "All human passion seems furtive and feverish beside this still rapture" ("Sun" 90). What's puzzling to her is that Woolf adds a further human/nonhuman animal comparison by turning to the fish in the aquarium tanks. She reports, "The fish are not explained. I've questioned a number of Virginia Woolf experts on this and no one appears to know why she adds fish to the lizards" (152). Her survey method falters at an instance of literary difficulty and perhaps an invitation to close reading.

Carson's answer, however, swerves from the text itself. She proposes Woolf was interested, in her essay as much as her double-date expedition, in triangulated desire and so introduces a complicating third element. Resolving her research question with extra-textually bolstered speculation and against-the-surface subtext, Carson evades Woolf's evident reason for turning to the fish, their encapsulation of the final argumentative stage in her extra-human reflections. But what's nonetheless interesting is that in veering away from Woolf's implied justification, Carson repeats a part of Woolf's essay arc, as though rather than teasing out the meaning of a pattern, overlapping with its surface becomes paramount. Just as Carson at the end of her essayette is intrigued by Woolf's associations' seemingly moving without reason, so Woolf is drawn to what the aquarium tanks reveal: the evolution of the forms of aquatic life without reason: "There the inhabitants perform for ever evolutions whose intricacy, because it has no reason, seems the more sublime." And she adds, "The most majestic of human evolutions seems feeble and fluctuating compared with the fishes'" (90). The lizards' "still rapture" stops time, but the fishes' bodies in all their complex variety perfect form, and do so in a way that confounds reason: "In their shape is their reason" (91). We, less suited to our environments and truly naked beneath our apparel, contemplate extinction, while they, "slipped into the world only to be themselves" (90–91), embody immortality. Woolf's immortality, whether embodied by the lizards' rapture that stills time or the fishes' form that belies our sense of evolution's teleology, might be seen as a deep and deeply nonhuman temporality, one that moves barely perceptibly, at once obscure, transparent. Both Carson and Woolf wish to pursue a force that motivates form yet moves with its own logic, each differently highlighting a technique of resonance more than reasoning. Woolf tracks the mind's eye; Carson follows but must unsee Woolf's apparent design to play up another that seems a subtext but can also be read as an overlay.

Stirner maintains that Benjamin thought of perception as the reading of surfaces not only in the case of the flâneur's feet treading urban ground but also with the ancient astrologer's "fleeting, transitory vision of the constellation of stars" (277). She cites his fragment "Doctrine of the Similar" in which Benjamin suggests, "the sphere of life which once seemed to be ruled by the law of similarity used to be much larger. It was the microcosm and the macrocosm" (65). The human capacity to perceive and produce similarity, "the mimetic faculty," seems to be declining, and for Benjamin, astrology is a case in point, since we are no longer capable of perceiving the "mimetic character" of constellations in a way that would allow us to comprehend a horoscope as "an original totality which astrological interpretation merely analyzed" or "the newborn" as endowed in an instant with "complete mimetic adaptation to the form of cosmic being." He likens the instant of birth to the flash of perception by which form is observed in stars' constellations and speculates that rapidity is crucial to the apprehension of similarity. Fittingly for us, he charts a three-part configuration in which the astrologer's apprehension of similarity is "like the addition of a third element, namely the astrologer, to the conjunction of two stars which must be grasped in an instant. Here the astronomer is cheated out of his reward, despite the sharpness of his observational tools" (66). The astronomer will need to learn to see, in effect, to become the strange connective tissue of space and time that renders an alignment a momentous, meaningful conjunction. Benjamin doesn't explicitly discuss the eclipse, yet from an astrological perspective it can be seen as a conjunction.[1] The child born at the moment of a total solar eclipse will have the sun and the moon in the same sign and house in an astrological chart and at very close degrees, merging their different energies and functions. The most useful astrologer might be the one who can make this the basis of the most sympathetic story.

And indeed, though we may have lost the intuitive feel for cosmic correspondence, Benjamin claims we have another "canon" that allows us to bring this "obscurity attached to the concept of non-sensuous similarity" to light: language (67). If in humanity's deep origins, "reading from the stars, entrails, and coincidences represented reading *per se*," writing might become "the most perfect archive of non-sensuous similarity," and forms of rapid reading, in which "similarities flash up fleetingly out of the stream of things only in order to become immediately engulfed again" might most rekindle this faculty (68), a faculty which is also, paradoxically, a source of literary difficulty, as seen in Woolf's logic of incongruity by which images drawn from "the queer pool in which we deposit our memories" will live on when oddly conjoined, while "The great sights will have died for lack of mates" (86). To scan for strange sparks of connectivity is to recognize that one reads in the dark. Alluding to Benjamin, Dipesh Chakrabarty has written that in the Anthropocene we might need to conceive "a new universal history of humans that flashes up in the moment of danger that is climate change" (221), but for Benjamin and Woolf the instantaneous flash was associated not just with emergency, but with birth, "original totality," sights' preservation, and with reading.

Note

1 I'm grateful here to Sarah Miles for her astrological knowledge and our exchange on eclipses.

Works cited

Bartkevicius, Jocelyn. "Thinking Back Through Our (Naturalist) Mother: Woolf, Dillard, and the Nature Essay." *Interdisciplinary Studies in Literature and Environment*, vol. 6, no. 1, 1999, pp. 41–50.

Benjamin, Walter. *The Arcades Project*. Translated by Howard Eiland and Kevin McLaughlin, Belknap Press of Harvard UP, 2003.

———. "Doctrine of the Similar." Translated by Knut Tarnowski, *New German Critique*, no. 17, 1979, pp. 65–69.

———. "The Work of Art in the Age of Its Technological Reproducibility: Second Version." *Selected Writings Volume 3: 1935–1938*, translated by Edmund Jephcott and Harry Zohn, edited by Howard Eiland and Michael W. Jennings, Belknap Press of Harvard UP, 2002, pp. 101–33.

Best, Stephen and Sharon Marcus. "Surface Reading: An Introduction." *Representations*, vol. 108, no. 1, 2009, pp. 1–21.

Carson, Anne. "Totality: The Color of Eclipse." *Decreation: Poetry, Essays, Opera*, Vintage, 2005, pp. 147–54.

Chakrabarty, Dipesh. "The Climate of History: Four Theses." *Critical Inquiry*, vol. 35, Winter 2009, pp. 197–222.

DiLeonardi, Sean Michael. "Cryptographic Reading: Machine Translation, the New Criticism, and Nabokov's *Pnin*." *Post45*, 17 January 2019, post45.research.yale.

edu/2019/01/cryptographic-reading-machine-translation-the-new-criticism-and-nabokovs-pnin/#.

Dillard, Annie. "Seeing." *The Abundance: Narrative Essays Old and New*, HarperCollins, 2016, pp. 151–72.

———. "Total Eclipse." *The Abundance: Narrative Essays Old and New*, HarperCollins, 2016, pp. 1–24.

Kimmerer, Robin Wall. *Gathering Moss: A Natural and Cultural History of Mosses*. Oregon State UP, 2003.

Millar, Liz. "Eclipse Photo." *Anthropocene: Arts of Living on a Damaged Planet*, Institute of the Arts and Sciences at the University of Santa Cruz, 2014, ias.ucsc.edu/events/anthropocene.

Moretti. Franco. "Conjectures on World Literature." *New Left Review*, no. 1, 2000, pp. 54–68.

Stirner, Simone. "A Technique of Closeness, an Art of Straying: Reading with Walter Benjamin." *New Literary History*, vol. 50, no. 2, 2019, pp. 271–91.

Tsing, Anna Lowenhaupt. *The Mushroom at the End of the World: On the Possibility of Life in Capitalist Ruins*. Princeton UP, 2015.

Tsing, Anna, Heather Swanson, Elaine Gan, and Nils Bubandt, editors. *Arts of Living on a Damaged Planet: Monsters of the Anthropocene*. U of Minnesota P, 2017.

Woolf, Virginia. "The Sun and the Fish." *The Crowded Dance of Modern Life: Selected Essays Volume 2*, edited by Rachel Bowlby, Penguin, 1993, pp. 86–91.

11

JOHN MASEFIELD'S "THE PASSING STRANGE"

Derangements of scale

Timothy Clark

Close reading, distant reading? The spatial terms are metaphors which I propose to consider through the concept of "scale," and in relation to what has been dubbed "scale critique." Scale critique is the intellectual discipline of studying how the nature of an entity, process, event, or the interpretation of a text transforms itself, even drastically, in relation to the varied temporal and spatial scales at which it is considered. It also considers critically the way assumptions about scale can be misleading or deceptive.

Recent ecocritical arguments have offered a new defense of literary representation, based the ability of literary texts to project and make acknowledged nonintuitive temporal or spatial frames. "Literature has always explored the tensions between the human mesoscale and different micro- and macroscales" (Pieter Vermeulen, 96). Vermeulen refers to such examples as Jonathan Swift's *Gulliver's Travels* (1726) or Olaf Stapledon's *Last and First Men (1930)* (96–97). My particular interest in this paper is in some poetic and rhetorical techniques which present human life on an expanded spatial or accelerated temporal scale, and the moral and other effects of this mode of representation. A poem by John Masefield (1878–1967) will offer a provocative focus.

"Scale critique" (a term proposed by Derek Woods) is an emerging critical practice in the environmental humanities. Thinkers in this field have found that to engage environmental issues of a dauntingly global nature often demands a reconsideration of assumptions about scale. This means developing the ability to read literary and other texts with much greater tolerance of cognitive dissonance, for what can read as coherent, welcome, and praiseworthy at one scale (increased prosperity, "improved" infrastructure, growing population) may be understood to be very different on another.[1]

Scale should be understood not just as a variant of perception but as a fundamental structure of human or in fact any imaginable entity. It is an ontological characteristic: a metal bridge of a tiny size has very different physical properties from one identical in shape but a hundred times bigger. Humans experience phenomena at a mostly fairly stable and consistent speed – too slow a perception would tend toward a static world, one in which nothing happens (the running of an animal would resemble the imperceptibly slow growth of plant), too fast, and everything would blur. We internalize this form of perception as a usually unconscious and deceptively self-evident norm. We understand distance, height, and breadth in terms of the given dimensionality of our embodied existence. The assumption of a seemingly "normal," embodied *scale* here acts as a kind of paradoxically originary representation. It projects or presents a specific version of the world, mediated through our bodily finitude and its particular dimensions. This may in turn become represented, and is reinforced by the familiar conventions of realism in art, photography, film, or literature, but also in nonrealist arts, whose effects of the fantastic or estrangement are still gauged by their felt distance from such embodied realism. Nevertheless, its status as a norm is merely a contingency. Defending the cognitive estrangements wrought by science-fiction, Seo-Young Chu proposes: "What most people call 'realism' – what some critics call 'mundane fiction' – is actually a 'weak' or low-intensity variety of science fiction, one that requires relatively little energy to accomplish its representational task insofar as its referents . . . are readily susceptible to representation" (7).

What, then, of scale in the so-called Anthropocene?[2] Modern humanity lives increasingly in a counterintuitive world whose reality does not conform to many assumed norms of previous decades. Whatever the merits or otherwise in the notion of an Anthropocene in the strict sense of a new geological epoch, it seems uncontroversial to think of it as at least a different *historical* category, one with which a recognition of the finitude of the earth itself becomes increasingly undeniable in a wide range of human affairs, in which we "remain perplexed as to how to find our place among phenomena, which are at once immensely vaster than we are, and yet subject to our affect" (Latour 93).

In effect, any clear distinction between what is "close" and what is "distant" has severely eroded. We must acknowledge that we now live in an environment in which old, once harmless norms, on expectations of personal space, resource use, and even the desirability of children, begin to acquire an "autoimmune," destructive quality. A person in the 2020s may well be living a life much like that of generations before, using energy in various ways, consuming food and resources, making waste, needing so much space, and so on. Nevertheless, across the invisible, blurred material or ecological threshold formed by the fact of so many other people now doing the same thing, and with more and more of the world dominated by programs of capitalist accumulation, what this person is actually doing has now become significantly different from people in the past, given the various and often opaque threshold and scale effects of the environment.

Scale and reading

Is there a "right" time frame by which to consider a text and how do different temporal scales affect the way something is understood? How, say, would innumerable twentieth- and twenty-first-century texts and films taking easy car and air travel as an unproblematic plot device read in the twenty-third century, hypothesized as a time of climate chaos and the mass displacement of populations? More pointedly, might even the most common-sensical realism of innumerable contemporary novels come to be read as merely limited or bad "science fiction," as Seo-Young Chu's argument suggests? The issue of representing human life artistically on the very large scale is now a common source of debate in relation to engaging with or imaging the so-called Anthropocene. Vermeulen states the issue in relation to debates about the novel and the challenges of representing the Anthropocene:

> If the novel form has traditionally been invested in the exploration of the fate of the individual and its relation to its social contexts, the discourses on the Anthropocene and on the geological ramifications of human culture . . . present the contemporary novel with a new challenge: that of scaling up its imagining of the human to the dimensions of biological and geological time.
> (Vermeulen (2015), 69)

The difficulty is making a representation at this scale that is intelligible, emotionally and aesthetically effective, without its being simplistic or too distorting. Derek Woods observes: "Writers and artists often imagine changes of scale as continuous expansion or contraction, or re-inscribe features of the human sensory world in scales at which they could not exist" (Woods (2017), 65). There is also the supplementary challenge of producing in an audience a genuine realization, if possible, that the "normal" scale of human bodily perception and day-to-day experience is no norm in any absolute sense, but is rather a biological contingency, even if it is also that to which the large and smaller scale representations or symbols must appeal if they are to be most felt. It is a case of using the immediacy of the "normal" scale in aesthetic techniques whose force is precisely to discredit that seeming norm (as when some material process in the long history of the earth is said to happen "geologically in the blink of an eye").

Although the critical focus has been overwhelmingly on novels[3] this challenge of scalar fluidity has arguably been met more routinely and forcibly in poetry. David Farrier observes "certain affinities between rhetorical manifestations of the Anthropocene and the kind of work we often expect a poem to do" (4–5). He continues:

> Poetry can compress vast acreages of meaning into a small compass or perform the kind of bold linkages that it would take reams of academic argument to plot; it can widen the aperture of our gaze or deposit us on the brink of transformation. In short, it can model an Anthropocenic perspective in which our sense of relationship and proximity (and from this, our ethics) is stretched and tested against the Anthropocene's warping effects.
> (5)

Incidentally, the relative brevity of much poetry must be another source of its value here, at a time when ecocriticism risks becoming boring by settling too easily into the limited and increasingly formulaic groove of expounding relevant and mostly recent novels.

The broader temporal or spatial scales imposed by thinking about the global environmental crisis, the so-called Anthropocene, can inform reconsideration of poems whose strategy is to highlight the finitude of the human scale through a pointedly accelerated pace of narration by, say, parsing into one sentence the process of a whole life. A paradigmatic instance might be the third line of Alfred Tennyson's "Tithonus:" "Man comes and tills the field and lies beneath" (112). Or there are those bizarre, powerful stanzas by Emily Dickinson that project a reader's consciousness into the experience of imaginary centuries after death ("Since then – 'tis centuries. . . .'" (350)). Another famous example would be that of Rainer Maria Rilke's "angels" in his *Duineser Elegien* (1923), superior, deathless beings whose exist within a different time scale in which human lives are only perceived indistinctly as blurred, fleeting mists.

The provocative effects of a speeded time frame are taken to an extreme, virtuoso degree in John Masefield's "The Passing Strange," a poem of 72 lines first published in *The Yale Review* in April 1920. The text is enigmatic to say the least, but it can safely be called a philosophical poem: its topic is human life viewed in totality as a scene of moral contradiction, of inconsistency, self-ignorance, possession by passion, and evanescence. "Passing Strange" is obviously not about the Anthropocene (to provisionally accept that term), but it forces its reader to engage with formal and epistemological issues that have emerged as crucial in debates on the global environmental predicament. This is the question of the effects, cognitive and ethical, of switching drastically between the scales at which one thinks.

The poem opens:

> Out of the earth to rest or range
> Perpetual in perpetual change,
> The unknown passing through the strange.
>
> Water and saltness held together
> To tread the dust and stand the weather,
> And plough the field and stretch the tether,
>
> To pass the wine-cup and be witty,
> Water the sands and build the city,
> Slaughter like devils and have pity,
>
> Be red with rage and pale with lust,
> Make beauty come, make peace, make trust,
> Water and saltness mixed with dust;
>
> . . .
> Fashion an altar for a rood,
> Defile a continent with blood,
> And watch a brother starve for food:

When Masefield wrote "The Passing Strange" he was known mainly as the author of darkly provocative long narrative poems ("The Everlasting Mercy" (1911), "The Widow in the Bye Street" (1912), and "Dauber" (1913) (Masefield, *The Collected Poems*, 87–131, 133–90, 191–245)). Paul Binding writes of this now largely forgotten or dismissed poet[4] that in his best work "we find an uncomfortable, troubled and troubling [vision], open to and outraged by pain and gratuitous suffering, offering us no tidied-up view of human nature either in its individual or social manifestations, particularly the last" (Binding 10). In "The Passing Strange" the tone is one of intense anti-mystification, even to the point of suggesting a latent if bitter identification with the violence described.

"The Passing Strange" sets itself off as a narrative of sorts, but its central protagonist is underdefined, enigmatic: who or what is it that appears "Out of the earth to rest or range?" This protagonist remains only minimally identifiable, being read presumably as a figure for human life in general, "unknown" in the sense that our own being and nature are a mystery to us philosophically. Later in the poem the human corpse is described as lying

> Down in the earth, in dark, alone
> A mockery of the ghost in bone,
> The strangeness, passing the unknown.

The encompassing generality of the text means that, if pressed, it falls *only just* short of being semantically empty. The two lines, "The unknown passing through the strange" (for life), and "The strangeness, passing the unknown" (for death) both resemble a sum of near negative numbers or of zeros – what is not known "passing" through, or beyond, what is not familiar. Since the topic seems to be being alive itself, the most familiar thing, then Masefield's poetic aim is one of maximum estrangement. The opening three lines, if one so wished, could even be parsed as an anticipation of Martin Heidegger's *Being and Time* (1927), on the nature of human existence as *Dasein,* or "being-there": "Ontically, of course, Dasein is not only close to us, – even that which is closest: we *are* it, each of us, we ourselves. In spite of this, or rather for just this reason, it is ontologically that which is farthest" (36).

"The unknown passing through the strange" is a provocative image of human lives. Strictly speaking *everything* cannot be "strange" as there is nothing else to compare it to, even though Masefield's title means, strictly, the "*surpassing* strange." The very abstracted nature of the verse enacts a shift from the ontic to the ontological. It is "strange" in the same way that human existence, the thing we know best ontically, is "unknown" ontologically. It also has no certain essence, being only "Perpetual in perpetual change."

Later in the poem, when this general everyman/woman protagonist is a dead body, seeping into the earth, it is called merely "the thing."

> Time will go by, that outlasts clocks
> Dawn in the thorps will rouse the cocks,

Sunset be glory on the rocks:

But it, the thing, will never heed
Even the rootling from the seed
Thrusting to suck it for its need.

"The Passing Strange" works at a level of abstraction that makes it provocatively if blurredly multi-scalar, however lucid its individual lines and clauses remain in themselves. The abstraction of the poem enables several possible readings of its striking scalar technique. If a key trope of "The Passing Strange" is read as one of speeding up, of an accelerated time scale, then one effect is that time that is usually invisible is become visible. Compare David Wood on the experience of looking at a tree: that if I look at a tree in front of me, what I see in that moment is only a misleading still, so to speak, of what a tree actually is. Wood writes:

> the life of the tree, the living tree, the tree of which we glimpse only a limb here, a trunk there, or views from various angles, this temporally extended persistent, growing tree, is invisible.
>
> (Wood (2005) 152)

Similarly, is human history, if one scales outward from the contingent limits of one's own immediate context, revealed as only what numerous writers and thinkers have long described? "History is a nightmare from which I am trying to awake" (Joyce, 34); "what experience and history teach is this – that people and governments never have learned anything from history, or acted on principles deduced from it" (Hegel, 6):

Fashion an altar for a rood,
Defile a continent with blood,
And watch a brother starve for food:

It could be a summary statement of the crimes committed in the name of religion. It might be the Spanish conquest of the Americas in three lines.

"The Passing Strange" takes up the broad, sweeping scale we find in historian Fernand Braudel's *longue durée*, or employed by practitioners of "distant reading" or ecocritics trying to evoke the perspective of "deep time." It evokes this sense of scale not with narrative expansion but through the conflation or collision of seemingly contradictory scenes or statements into the same space. Here human behavior itself, at its worst as well its best, becomes as repetitive as the cycles of day and night.

This is an example of how the sense of living in a new historical epoch can change readings of past literature, making Masefield's text suddenly of renewed interest, both in its topic and as example of the kinds of rhetorical and formal invention which that topic calls forth. "The Passing Strange," if immediately foreign to the "Anthropocene" debate, reinforces arguments that it is in the resources of poetry, with its demands of close reading, that some of the scalar challenges of representation can be most effectively met. It deploys linguistic structures whose formal patterning and collisions suggest dark correlations and determinations at work across, and, as it were, beneath, the individual lines and statements of the text. The use of three near identical rhymes in a row gives a sense of both an impersonal process and a mildly jingling sense of absurdity (e.g., "rood . . . blood . . . food;" "witty . . . city . . . pity"). "Slaughter like devils and have pity" enacts the same compression in the collision even between the first and second part of the one clause. The opening section of the poem mingles a sense of absurd horror with the emptiness of déjà *vu*.

The discordances are like deliberately arranged collisions of sense, a provocative dissonance, a refusal of "the unity of experience itself as man knows it in his own experience" (194), to cite the "new critic" Cleanth Brooks's defense of the effects of his well-known ideals of poetry and "close reading." Masefield's technique is germane to the more recent association of "close reading" with the analytical practices of deconstruction, as in Sara Guyer's study of John Clare which defends the "difficulty and discomfort of close reading" against "those who have dismissed it as expired, extinct, and belonging to another era" (7). What Guyer means by close reading is the kind of "rhetorical reading" associated with the work of Paul de Man, with its close attention to the discordant, and the intellectually unassimilable. Close reading is "uncomfortable reading" (10).[5] Similarly, in Julian Wolfreys's *Reading: Acts of Close Reading in Literary Theory* (2000), "close reading" means reading alert to the aporetic, resistant to the temptation of reading and concluding prematurely. "Close reading" in Guyer's deManian sense teases out seeming discordances and aporias in often seemingly coherent discourse, while Wolfreys calls for acts of finer, unhasty discrimination (vii). In "The Passing Strange," however, there is no seemingly coherent surface to queston. It is the possibility of discerning ultimately meaningful moral differences that is being eroded, as if societies and histories across the planet, with their multiple achievements and horrors, were collapsing upon themselves into a sort of ethical and conceptual black hole.

One can read "The Passing Strange" in terms of disjunctions in time, that a community that enjoys prosperity and forms of "enlightened" equity at one time can become at another violent and murderous. Alternatively, it can be read in terms of disjunctions in space, that the peace and prosperity of the one group depends upon the brutal treatment of another. It could also express a grim truth depicted in Masefield's longer narrative poems, the easily evaded fact that most human groups actively enjoy denigrating others, and will seize upon occasions to do so, even in peace and prosperity.

Such a speeded or conflated narrative is profoundly dehumanizing, producing the cognitive discomfort of an impersonal dynamic. The poem reads as an extreme example of the way in which what seems right or wrong about any one event or situation can shift, or even be inverted, if the temporal or spatial context for it is altered. Categories of approval or disapproval would be only simplistic "stills," so to speak, of a process that is normally invisible at our day-to-day time scale. Relating this to our own twenty-first-century context, environmental problems are a particularly strong example of the kind of reversal/conflation of good and bad – something good in the immediate present (fertilize the fields) is bad over a longer time frame (pollution, eutrophication of rivers), a wider distribution of material spoils in the short term can mean severe resource depletion and increased poverty in the longer term. Such issues become morally fraught and controversial with the issue of human population in particular – a new beautiful baby/another mouth to feed.[6]

In "The Passing Strange" any "hermeneutics of suspicion" that a critic might apply to demystify seemingly benign social behavior can seem otiose, for here that behavior is immediately juxtaposed with murder:

> To pass the wine-cup and be witty,
> Water the sands and build the city,
> Slaughter like devils and have pity,

One brief line effectively hollows out, mocks the sense of adequacy or normality in another. The lines set up multiple interference effects with each other whose tone is very hard to describe, being disconcertingly light, even at first to a nihilistic effect. There seem to be here no illusions to unmask.

This is one instance of what is termed "unnatural" narrative in the sense of not fitting the supposedly natural model of "experientiality," that is, "the quasi-mimetic evocation of 'real-life experience'", which "correlates with the evocation of consciousness" and "reflects a cognitive schema of embodiedness that relates to human existence and human concerns" (Jon Hegglund, 27–28). "The Passing Strange" brings a peculiar pressure to bear on the overly normative force of supposedly "natural" mimesis by changing its pace and scope, and by keeping always open the question of whether the movement from one clause, or line, or three-line stanza to the next is projecting any kind of narrative at all in the sense of linearity and causal connection. To change the spatial or temporal scale at which an object is considered opens up its context in ways that can reveal previously secure determinations as no more than phantasms. While ecocriticism would make such a point through the analytical paraphrase of the plot or characters from some relevant novel, Masefield's few lines work with the force of an immediate, rhetorically aggressive shock.[7]

In this respect "The Passing Strange" underlines what Claire Colebrook and others have argued in relation to humanist responses to the so-called "Anthropocene," and its implicit demand to submit human life to forms of radical new questioning. This is to refuse widespread notions that the crisis calls for some return to a supposedly more connected, coherently moral, or "ecological" form of human being. Rather, it is the revelation of something that has always been in operation, of the human as part of a biological and ecological dynamic in which destruction and extinction, including self-extinction, have always been inherent, even constitutive and elemental.[8]

The grand scale and moral accounting

Let us return to the issue with which this chapter opened that of the discrepancy between the reality being called the Anthropocene and the scalar reality of "normal" personhood and the human sensorium. How far can scales of human perception be altered? "It is precisely those spaces, phenomena, and experiences which seem most 'natural' to us that are deserving of critical deconstruction as historically constituted" (Shane Gunster 221). Our sense of scalar norms may be altered by technology (Rome is now much "nearer" Paris than in the nineteenth century, a person living in Australia is not out of communicative reach of someone in London), and what is experienced as close or distant is clearly a matter of historical construction to a significant degree. Nevertheless, there is still a dangerous mismatch between the spatial givens of individual personhood and the contemporary context. David Wood writes:

> Our palette of affective responses is the product of evolutionary history, in which individual and group reproductive success and survival have been the selective mechanisms. But the circumstances in which our passions organize themselves to promote survival and flourishing have changed.
>
> (Wood (2019), 126)

A deeper issue may be the "manifest image" (Wilfrid Sellars) which we each have of our personhood or subjectivity, of *what being a person means*, that commonsense philosophy of mind according to which human beings are relatively free agents and that what they do can be explained by reference to such entities as inner thoughts, decisions, desires, projects, and intentions, each held to direct meaningful sequences of action. This commonsense model of mind evolved over millennia as seeming to provide a reliable framework in which to conceive of ourselves and our interactions. However, what if one emergent effect of the Anthropocene were an end or, at the very least, the severe erosion of the credibility of this seemingly self-evident image of the human (one which a dominant tradition of realism in the novel has reinforced)? It is also, in retrospect, one of the fundamental questions raised by "The Passing Strange."

One effect of Masefield's technique of conflation is that the protagonist of "The Passing Strange" is a disturbingly conflicted kind of every man or woman. The poem deliberately blurs the normal, seemingly comfortable distinction between the one and the many. This is the result of the way Masefield highlights actions (raising an altar, murdering people, admiring beauty, becoming "red with rage and pale with lust"); in no place is it said who, specifically, performs these actions. The effect is of a broadly universal/impersonal humanity as the agent, coming "out of the earth to rest or range" and living out repeated and disturbing moral contradictions.

For Masefield, a sense of the illusory and fragile nature of the bounds of personhood took the form of an interest in doctrines of reincarnation. He was attracted to broadly Buddhist doctrines of human existence as a process of reincarnations, as he stated in 1952:

> Many have believed that the spirit of the human being lives many lives while seeking release from the chains of desire. It may be so. Recurrent world wars and the terrible peaces that follow them must have loosed some hundreds of millions from all chains of desire for any more life after this: perhaps some grace of remission of sentence may be granted to modern Europeans.
>
> (152)

Masefield was attracted, without dogmatic commitment, to a broadly Buddhist notion of human existence as a cycle of reincarnations (*samsara*). This suggests a further possible reading of the scalar conflation in "The Passing Strange." Perhaps a reader could refer to Buddhism's "four noble truths", that all existence is suffering, and that this is partly caused by ineradicable desire of something lacking? Masefield's image of human life is a drama of possession and transformation by desire, as fraught but also bringing kindness and peace, whether living "as a woman or a man":

> Love like a madman, shaking, blind,
> Till self is burnt into a kind
> Possession of another mind;
>
> Brood upon beauty, till the grace
> Of beauty with the holy face
> Brings peace into the bitter place;
>
> Prove in the lifeless granites, scan
> The stars for hope, for guide, for plan;
> Live as a woman or a man;
>
> Fasten to lover or to friend,
> Until the heart break at the end
> The break of death that cannot mend.

One attraction of Buddhism is that it offers a broad scheme of moral accounting in human life over large time scales, with reincarnation entailing a process in which past actions influence future lives and incarnations. For instance, David Mitchell's novel *Cloud Atlas* of 2004 adapts the reversioning of one character in another over many generations to express contemporary environmental concerns about intergenerational ethics.[9]

Mitchell would exemplify the kind of ethical causality at work in the notion of *Karma*, the sum future effect of a life's moral actions. An earlier poem by Masefield entitled "A Creed" exploits Buddhist notions as just such a moral accounting system, tinged even with an incongruous individualism ("This hand, this hand that holds the pen,/Has many a hundred times been dust/And turned, as dust, to dust again. . . . My road shall be the road I made;/All that I gave shall be repaid"). In "The Passing Strange," however, the stress is on the disturbing disjunctions between actions/lives, in a way that seems to mock any sort of ethical accounting of cause and effect. What might be readable, just, in terms, of a variant Buddhist doctrine of one and diverse lives at various times (as in "A Creed") remains a statement of humanity en masse as morally chaotic and ethically opaque. By contrast, the Buddhist moral system, even as adapted by Mitchell in relation to global environmental collapse, is a homogenous scalar extrapolation of what is familiarly good or bad, one not open to the kinds of discontinuity inherent in scale effects. Rather than reflect any form of cosmic moral accounting system, "The Passing Strange" is closer to the philosopher David Wood when he writes:

> Modern culture bears pervasive witness to the breakdown in the traditional lived experience of time. We may not, literally, have lost our memories, but loss still attaches itself to memory – and not just to specific items of the past, but to the whole framework in which past, present, and future are coherently connected.
>
> (Wood (2007), 10)

Earth processes

One effect of refusing notions of consistency and linear narrative of the kind associated with our immediate, illusory sense of personhood is a sense of entrapment in the arbitrary, in what does not "make sense."

What sense is made of the large-scale view projected in the poem is achieved by Masefield's embedding of the contradictions of human life within larger natural or cosmological processes. Masefield's scalar overview of human life diminishes even its most seemingly praiseworthy or endearing aspects: compassion, the quest for knowledge, and love. These now seem only parts of a demeaning and predictable syndrome of contradictory behaviors, expressible in tiny phrases as if merely two sides of the same coin ("Slaughter like devils and have pity"). Yet, expanding the scale of perception also works to contradictory effect. "The Passing Strange" exalts the possessor or reader of its expanded knowledge beyond the more customary borders of human finitude, like a view of human society through the wrong end of the telescope, or the sight of the earth from space.

One can say that the poem is about the earth, as the impersonal, non-meaning foundation and medium of human lives. The human life process, to give a provisional name for Masefield's protagonist, comes "Out of the earth" (line 1) as a compound that perhaps recalls the inveterate notion (actually a fallacy) that human blood retains the makeup of sea water: "water and saltness held together."

It then later returns to the earth: "Then to lie useless, helpless, still, /Down in the earth," becoming only part of the physical processes there.

Consider the text's various images of dust, of dust, salt and water as compounds that make up the living human body ("Water and saltness mixed with dust;" "this water mixed with dust"). There is both dust as the earth, the material beneath our feet which cultures mold and exploit and feed from:

> Water and saltness held together
> To tread the dust and stand the weather,
> And plough the field and stretch the tether

And there is also dust as that to which the human body returns as the human generations and the revolutions of the earth continue:

> The corn is sown again, it grows;
> The stars burn out, the darkness goes;
> The rhythms change, they do not close.
>
> They change, and we, who pass like foam,
> Like dust blown through the streets of Rome,
> Change ever, too; we have no home [. . .]

This identification with the earth underlies the final, fantastical passages of speeded-up time in which the dead material still participates:

> Only a beauty, only a power,
> Sad in the fruit, bright in the flower,
> Endlessly erring for its hour . . .

The "earth" informs Masefield's image of human life in terms of transience and lack of essence. Its impurity is a refusal of the idea of some simple nature or origin – some lost essential nature from which evils such as violence or corruption are a falling off, refusing metaphysical or religious conceptions in which the story is told of how some original presence or simplicity came to be contaminated or divided by complexity.[10] "Perpetual in perpetual change,/The unknown passing through the strange. /Water and saltness held together." The "erring" seems to be fundamental, and a process without purpose, for its flower is "bright" but the subsequent fruit, the biological aim of a flower, is only "sad," even as the poem ends abruptly with a kind of Nietzschean acceptance of or identification with the whole cyclical, imperfect happening of "Life" as affording a kind of "joy."

These lines about the earth convey here a surprising contemporary sense of life and death as inextricably connected in the one economy or process, expressed by the way Masefield conflates earth, associated with growth and fertility, with dust, associated with death. The two are just different phases of the same thing.

The human body is compounded of dust, salt, and water. Death is not the opposite of life, but its dynamic, even its constitutive grammar so to speak.

Repercussions?

Masefield's techniques of cognitive dissonance acquire a new interest in our own overwhelmingly conflicted and complex global predicament, in which, as Braden Allenby and Daniel Sarewitz write, "If simplification is a necessary function for cognition and for day-to-day functioning, it is nonetheless problematic if the essence of the environment is, in fact, its complexity" (120). Any moral authenticity in human life must entail accepting the limits of any self-knowledge, given that we live in a world that may be beyond the reach of rational enquiry or the principle of noncontradiction. Allenby and Sarewitz defend an enlightened conception of "muddling," though not in the sense of relativism:

> [W]e mean muddling as in understanding that ethics itself is an evolving system in a rapidly changing world. (It used to be the case, for example, that some people were not considered persons under some ethical and legal frameworks, a position that was strongly defended by people considered ethical at the time) . . . ethics . . . becomes partial as it becomes coherent – an ethical uncertainty principle that we cannot escape.
>
> (183)

If "close reading" has most often been associated with the reading of poetry, forms of modernist and postmodernist poetics must also now stand to be challenged. Take, for example, the philosopher Simon Critchley's celebration of Wallace Stevens, which opens with a statement of poetic principles. Critchley refers to Steven's well-known poem "Anecdote of the Jar," a meta-poetic text about how art transforms individual perception:

> I placed a jar in Tennessee,
> And round it was, upon a hill.
> It made the slovenly wilderness
> Surround that hill.
>
> The wilderness rose up to it,
> And sprawled around, no longer wild.

Critchley writes: "Poetry allows us to see things *as they are*. It lets us see particulars, being various. . . . Things change when the poet sings them, but they are still our things: recognizable, common, near, low" (11, emphasis added). However, this statement is still based on the presupposition that the true aim of poetic perception and reading is the finite perception of an individual, at the most familiarly "given" scales of space and time. Critchley and Stevens both presuppose a phenomenological standpoint on particulars, but to frame that particular (say a jar, or a ship) in relation to far larger, unseen scales of space and time is a much greater challenge.

Do readings based on changing scale have some sort of redemptive function in the way we think, which is how most ecocritics tend to read texts inculcating a sense of "deep" or deeper time, or do the insights obtained lead only to forms of resignation? "The Passing Strange" might seem to fit the latter response. Nevertheless, as an exercise in scalar defamiliarization, it feeds positively into current debates.

In his *Nihil Unbound* Ray Brassier defends a version of nihilism against crude notions of it as the view that nothing matters, that human life is worthless, and so on: rather, "[t]he disenchantment of the world is to be celebrated as an achievement of intellectual maturity not bewailed as a debilitating impoverishment" (xi):

> Philosophers would do well to desist from issuing any further injunctions about the need to re-establish the meaningfulness of existence, the purposefulness of life, or mend the shattered concord between man and nature. Philosophy should strive to be more than a sop to the pathetic twinge of human self-esteem. Nihilism is not an existential quandary but a speculative opportunity.
>
> (xi)

"The Passing Strange" refuses any stance of moral or intellectual complacency (for what seems virtuous may always be the flip side of evil) and, until its speculative ending (perhaps), it refuses any sort of religious or transcendental overview that gives some redemptive or purposeful meaning to its processes and repeated scenes. What emerges is a notion of "close reading" as a mode of attention that resists premature synthesis, which recognizes ethical irresolution in refusing the complacency of the good conscience. It is not that Masefield somehow anticipated the Anthropocene and the way old norms about human space, time, and responsibility have now collapsed, along with any sustainable distinction between what is "close" and what is "distant." It is rather that to read "The Passing Strange" now is to be struck by a chastening sense of human existence that our contemporary context of overreach makes more starkly visible, and harder to evade with premature schemes of resolution.

Notes

1 This is the central argument of my *Ecocriticism on the Edge*. Dipesh Chakrabarty also writes of climate change and the now widely recognized "need to think human agency over multiple and incommensurable scales at once" (Chakrabarty (2012) 1).

2 For an argument for caution of this over-used term see my *The Value of Ecocriticism* (43–49).

3 See, for instance, Mahlu Mertens and Stef Craps 2018.

4 "If ever there could be said to be a corpse among twentieth-century writers of stature, Masefield is as good a candidate as any" (Binding 6).

5 Guyer is responding to the current vogue of "distant reading" associated with the work of Franco Moretti. This is reading that does not minutely analyze individual texts but which takes on board large corpora, even statistically, often to challenge inherited assumptions about genre, taste or period in literary studies, assumptions shown to be based on a very limited and selective body of the texts (the "canon" for instance).

6 Chakrabarty observes such a split, remembering "the pride with which today the most ordinary and poor Indian citizen now possesses his or her smart phone or a fake and cheap substitute. The lurch into the Anthropocene has been globally the story of some long-anticipated social justice as well, at least in the sphere of consumption.

This justice between humans, however, comes at a price. The result of growing human consumption has been a near-complete human appropriation of the biosphere" (Chakrabarty (2015) 52).

7 Marco Caracciolo, following Karin Kukkonen, gives a useful summary of the four distinct dimensions that normally make up a narrative "plot." "The Passing Strange" has a questioning relation to all of them.

Temporal relations between events (i.e., what comes before what); causal relations between characters and events (i.e., who performs what action and for what purpose); thematic coherence (i.e., what the story is about and what function or "point" it has within the larger communicative act); and affective dynamics (i.e., why the story is interesting and emotionally satisfying) (48).

8 See, for example, Lynes and Colebrook.

9 Mitchell enables:

an alternative approach to linear temporality, whose spiraling cyclicality warns of the dangers of seeing past actions as separate from future consequences, and whose focus on cross-temporal interconnection demonstrates the importance of collective, intergenerational ethical action in the face of ecological and humanitarian crises. (Rose Harris-Birtill "'Looking down time's telescope at myself': reincarnation and global futures in David Mitchell's fictional worlds," pre-submission draft, https://risweb.st-andrews.ac.uk/portal/en/researchoutput/looking-down-times-telescope-at-myself-reincarnation-and-global-futures-in-david-mitchells-fictional-worlds(6e82e2b3-ac34-4a84-9e4a-56beb558e4e6).html)

10 If, as "[Jacques] Derrida says, there is complexity at the origin . . . *the origin is not simple*" (Bennington 231).

Works cited

Allenby, B.R. and Sarewitz, D. *The Techno-Human Condition*. MIT Press, 2011.

Bennington, Geoffrey. "Foundations." *Textual Practice*, vol. 21, no. 2, 2007, 231–49.

Binding, Paul. *An Endless Quiet Valley: A Reappraisal of John Masefield*. Logaston Press, 1998.

Brassier, Ray. *Nihil Unbound: Enlightenment and Extinction*. Palgrave Macmillan, 2007.

Braudel, Fernand. *On History*. Trans. Sarah Matthews. University of Chicago Press, 1980.

Brooks, Cleanth. *The Well Wrought Urn*. Harcourt Brace, 1947.

Caracciolo, Marco. "Object-Oriented Plotting and Nonhuman Realities in DeLillo's *Underworld* and Iñárritu's *Babel*." *Environment and Narrative: New Directions in Econarratology*. Eds. Erin James and Eric Morel. Ohio State University Press, 2020, 45–64.

Chakrabarty, Dipesh. "The Anthropocene and the Convergence of Histories." *The Anthropocene and the Global Environmental Crisis: Rethinking Modernity in a New Epoch*. Eds. Clive Hamilton, Christophe Bonneuil, and François Gemenne. Routledge, 2015, 44–56.

———. "Postcolonial Studies and the Challenge of Climate Change." *New Literary History*, vol. 43, no. 2, 2012, 1–18.

Chu, Seo-Young. *Do Metaphors Dream of Literal Sleep: A Science-Fictional Theory of Representation*. Harvard University Press, 2010.

Clark, Timothy. *Ecocriticism on the Edge: The Anthropocene as a Threshold Concept*. Bloomsbury, 2015.
———. *The Value of Ecocriticism*. Cambridge University Press, 2019.
Colebrook, Claire. "Extinguishing Ability: How We Became Postextinction Persons." *Eco-Deconstruction: Derrida and Environmental Philosophy*. Eds. Matthias Fritsch, Philippe Lynes, and David Wood. Fordham University Press, 2018, 261–76.
Critchley, Simon. *Philosophy in the Poetry of Wallace Stevens*. Routledge, 2005.
Dickinson, Emily. "Because I Could Not Stop for Death." Poem 512 in *Emily Dickinson: The Complete Poems*. Ed. Thomas H. Johnson. Faber, 1970, 350.
Farrier, David. *Anthropocene Poetics: Deep Time, Sacrifice Zones, and Extinction*. University of Minnesota Press, 2019, 4–5.
Gunster, Shane. "Fear and the Unknown: Nature, Culture, and the Limits of Reason." *Critical Ecologies: The Frankfurt School and Contemporary Environmental Crises*. Ed. Andrew Biro. University of Toronto Press, 2011, 206–28.
Guyer, Sara. *Reading with John Clare: Biopoetics, Sovereignty, Romanticism*. Fordham University Press, 2015.
Hegel, Georg Wilhelm Friedrich. *Lectures on the Philosophy of History* [1837]. Trans. J. Sibree. Dover Publications, 1956.
Hegglund, Jon. "Unnatural Narratology and Weird Realism in Jeff VanderMeer's *Annihilation*." *Environment and Narrative: New Directions in Econarratology*. Eds. Erin James and Eric Morel. Ohio State University Press, 2020, 27–44.
Heidegger, Martin. *Being and Time*. Trans. John Macquarie and Edward Robinson. Basil Blackwell, 1980.
Joyce, James. *Ulysses* [1922]. Random House, 1961.
Latour, Bruno. "Anti-Zoom." *Scale in Literature and Culture*. Eds. Michael Tavel Clarke and David Wittenberg. Palgrave Macmillan, 2017, 93–101.
Lynes, Philippe. "The Posthuman Promise of the Earth." *Eco-Deconstruction: Derrida and Environmental Philosophy*. Eds. Matthias Fritsch, Philippe Lynes, and David Wood. Fordham University Press, 2018, 101–20.
Masefield, John. *The Collected Poems of John Masefield*. William Heinemann, 1924, 663–65.
———. "A Creed." *Ballads and Poems*. Ed. Elkin Mathews, 1910, 64–66.
———. "The Passing Strange." *The Yale Review*, vol. IX, April 1920, no. 3, 449–51.
———. *So Long to Learn: Chapters of an Autobiography*. William Heinemann, 1952.
Mertens, Mahlu and Craps, Stef. "Contemporary Fiction Vs. The Challenge of Imagining the Timescale of Climate Change." *Studies in the Novel*, vol. 50, 2018, 134–53.
Mitchell, David. *Cloud Atlas*. Sceptre, 2004.
Rilke, Rainer Maria. *Duineser Elegien* [1923]. Ed. Joseph Kiermeier-Debre (Bibliothek der Erstausgabe). Deutscher Taschenbuch, 1997.
Sellars, Wilfred. "Philosophy and the Scientific Image of Man." *Frontiers of Science and Philosophy*. Ed. Robert Colodny. University of Pittsburgh Press, 1962, 35–78.
Tennyson, Alfred. "Tithonus." *The Poems of Tennyson*. Ed. Christopher Ricks. Longman, 1968, 112–18.
Vermeulen, Pieter. "Don DeLillo's Point Omega, the Anthropocene, and the Scales of Literature." *Studia Neophilologica*, vol. 87, 2015, 68–81.
———. *Literature and the Anthropocene*. Routledge, 2020.
Wood, David. *Deep Time/Dark Times: On Being Geologically Human*. Fordham University Press, 2019.
———. *The Step Back: Ethics and Politics after Deconstruction*. SUNY Press, 2005.

12

FROM SCALE TO ANTAGONISM

Reading the human in Kurt Vonnegut's *Galápagos*

C. Parker Krieg

Against the double erasure

The Anthropocene calls for practices of reading that rescue the human from the double erasure of frameworks that elide social difference under the figure of the *anthropos* and, at the same time, bury the *anthropos* beneath geological, evolutionary, and atmospheric forces. To redress this double erasure, this chapter moves from questions of scale to those of antagonism. To move from scale to antagonism means to articulate narratives within the contexts of cultural production, histories of the nation-state, and theoretical developments in the neoliberal university. Understood as that which "structures in advance *the very terrain* on which the multitude of particular contents fight for hegemony," antagonism overdetermines projections of scale and their narrative uses (Žižek 320). The Anthropocene is a totality structured by an antagonism between capitalism and the earth, which is reproduced through dispossession, exploitation, and exclusion. Where posthumanist readings, like Giorgio Agamben's, hope to "render inoperative the machine that governs our conception of man," a focus on the *political* contexts of literature better attends to the work of culture in an ecological crisis whose genesis and outcomes are deeply social (92). This chapter takes up Kurt Vonnegut's 1985 novel, *Galápagos*, a text that has lent itself to considerations of scale that eclipse the human, and reads it in the context of the author's criticism of war, capitalism, and the complicity of professional class intellectuals in planetary crises. It considers the novel's satirical evolutionary narrative in relation to recent moves "beyond" the human, either to post-historical animality or digital disembodiment, and brings together critics who suggest that these are false solutions. I argue that close reading in the Anthropocene must not reproduce sublime derangements of scale that symptomatically reinforce the double erasure, but rather connect literature to social antagonisms that reproduce the Anthropocene as a common condition.

To do this, I propose a deeper engagement with practices and outlooks developed by cultural studies, taking Simon During's description of close reading as an "assemblage of practices" as a starting point (85). These practices include *interpretation*, or an account of meaning in a text that differentiates itself from other accounts; *contextualization*, or descriptions of the historical conditions of a text's production; *literary scholarship*, which situates or redefines a text's relationship within the history of literary texts; and *evaluation*, which assesses the value of a text on the basis of particular criteria (e.g., moral, political, and ecological) These overlapping practices emphasize connectivity and influence, critical reception, and invented tradition. Moreover, they position the reader against ahistorical moves that isolate or withdraw an object in its being apart from its constitutive relations. To this, I add *articulation*. As developed by Stuart Hall, articulation – in the sense of linking – connects authors, texts, readers, and critics, each of whose activity is understood as taking place alongside the others across overlapping political contexts that determine its meaning.[1] In *We Gotta Get Out of This Place*, Lawrence Grossberg writes that "articulation links this practice to that effect, this text to that meaning, this meaning to that reality, this experience to those politics." Because, "these links are themselves articulated into larger structures," the critic effectively "reworks the context into which practices are inserted" (54). If this sounds knotty, it is because this locates reading the scale of social terrain where antagonism and meaning-making occur. By understanding close reading as a set of practices, this approach avoids the mimetic trap of (mis)taking a particular environmental image or metaphor as a "more ecological" model for reading. Just as greenwashing can hide unsustainable practices, so too must criticism not fall prey to allegorical readings based on abstract homologies between ecological or geological models and the production of culture. Rather, the task of reading is to connect both textual objects and interpretive practices to their institutional contexts and political–economic conjunctures, in order to identify emerging formations of ecological publics.

Whatever form close reading takes in the Anthropocene will be connected to the status of the humanities in the university, but it need not begin nor end there. For example, when Marxism entered the university in response to the Great Depression and the rise of international fascism, it became a means for rereading U.S. literary culture in the interests of a large, multiracial, and industrial democracy. In the 1940s, the close reading of the New Critics depoliticized literature in favor of esthetic values. By rejecting both the "genetic" and "receptionist" readings of social analysis, this version instead "distances" and "depersonalizes," so as to interpret the isolated poetic object as a complex unity that exemplifies a corresponding complexity in reality itself (Leitch 25). While the microcosm of the university allowed students at state schools and ivy leagues alike to approach texts on the same footing, this practice participates in the general professionalization of a "new class" of intellectuals, divorced from a public ethos (Schryer 663). Understood economically, New Criticism is a "poetics of Fordism," in which close reading as a mode of consumption intensifies to become a mode of production through interpretation (Tumino 172; Nealon 172). This change is also registered in the communicative labor and network models of the post-Fordist workplace (Boltanski and Chiapello 143). The rise of immaterial labor and postmodern consumer ethos encourages generative readings from pluralized viewpoints, while the interpretive commitments of readers are opened up by ethnic minority, feminist, LGBTQ, and environmental critics, informed by theories of productive subjectivity. Each has entered and adapted with the institutional history of the university.

Environmental humanists are quick to observe the possibilities and limits of the Anthropocene as "a millennial concept" after the supposed end of history and grand narratives (Menely and Taylor 5). For Bonneuil and Fressoz, a new class of "anthropoceneologists" has emerged, led by thinkers like Bruno Latour and Ulrich Beck, who promise to "make everything new" with accounts of reflexive modernity that overlook how deeply reflexive past societies (both ancient and modern) already were about the environment (2016). Anthropoceneologists offer a narrative break with the bad modernity of the past, *this time* guided by scientists (redefined as cultural experts) to preside over a domesticated space of global hybridity and risk. This does not break with modernity so much as reenact the break within modernity itself, by positing a time of static distinctions interrupted by dissolvent fluidity. It simply *is* the experience of modernity only updated for contemporary production. This transforms struggles for political control over the means by which society reproduces itself into problems of knowledge. Moving within this zeitgeist are activist-oriented speculative readings that curiously produce *unknowing* and *undecideability* as the desired outcome. Against mounting projections of diminishing planetary prospects, these readings prioritize open-endedness, and the chance to affectively reconstitute what Bifo calls *futurability* or what Donna Haraway calls speculative fabulation, as an existential claim on the future. Some even describe this scholarship as *science fiction* which produces a future to reckon with (Swanson, Bubandt, and Tsing 2015). This is understandable. But if these approaches are guilty of overinflating individual or scholarly agency, the opposite holds for posthuman media theory that views the displacement of the human (intellect and desire) by techno-capital as a radical transgression. Combining speculation and eschatology in the "critical climate change" series, Tom Cohen claims there was never a human to think beyond (9). These positions prefigure the very dissolution they predict, and dissolve along with it an understanding of the human as the species being that both makes its culture and regenerates it. Indeed, the conflicting accounts and scales of the Anthropocene offer a new grand narrative for interdisciplinary legitimation, while the debate over the *anthropos* restages mid-century "crisis of Man" discourse on racism and technology after globalization (Toivanen et al. 2017; Greif 2015).

Against scales that retroactively naturalize capitalism and settler-colonialism, scholars assert the excluded cultural histories of environmental violence they contain (Lewis and Maslin 2015; Moore 2017; Whyte 2017). These efforts have more in common with Edward Said's "return to philology" than with the theoretical antihumanism from which he distanced himself. For Said, humanism is a practice of reading that excavates "testimony that doesn't make it onto the reports;" it asks "whether an overexploited environment, sustainable small economies and small nations, and marginalized peoples outside as well as inside the maw of the metropolitan center can survive the grinding down and flattening out and displacement that are such prominent features of globalization" (549). Building on Said's humanism, Timothy Brennan critiques a set of emerging positions that displace this reading subject. In *Borrowed Light: Vico, Hegel, and the Colonies*, he outlines the depoliticizing effects of this "imperial posthumanism":

> It gives alienation a philosophical and scientific respectability: to disembody human skill and intelligence, to de-realize human will and effort, to unthink the human. The human subject alienated from his or her powers of cognition, divorced from the body, delivered from the uniqueness of species, and deprived of its will is the form of its sublime.
>
> *(232)*

In Brennan's expansive philology,[2] this posthumanism represents "a massively willed and historically determined effort to be done with will and history as human making" (232). Under the guise of challenging the rationalism of bourgeois modernity implicit in the Anthropocene, the "demotion of the human" paradoxically reinscribes the framework of the naturalistic sciences (233). "While it is true that market ideology relies on an active subject," Brennan argues, "the history of Western thought is marked throughout by a desire to cast the human within an indifferent nature as one of its expendable features, to blur its boundaries and repossess for the human being the image of his or her own bestiality and insignificance" (233). In much posthumanism, this disavowed scientism is joined to the cultural theories of interwar European reactionaries against the supposedly bad universalism of left thinkers, despite the significance of the latter in a century of anti-colonial movements. This bears on Brennan's account of the neoliberal university, where the subject that is remade through interpretive struggles over national narratives, popular texts, and world literature, faces displacement on two fronts: first, by well-funded digital humanities initiatives that substitute computation for the development of human capabilities; and then by the redefinition of agency through the nonhuman under the auspices of challenging a philosophical image of the human. In both, the exhaustion with "critique" can be read in conjunction with a "business climate and media culture that assumes the supremacy of the natural sciences," which strives to assimilate the humanities under its "managerial wing" (225). As it advocates for more-than-human sensibilities, the environmental humanities might, where it does not do so already, interrogate the complicities of this demoted human in its own Anthropocene reading practices.

For example, the philosopher Bernard Stiegler reconstitutes the *anthropos* through an "originary technicity" whose political nature is first identified by Plato in the *Phaedrus*. It is precisely because the *anthropos* is a creature who is individuated through inscription that it is vulnerable when supplements are subsumed by economic drives. For Stiegler, the Anthropocene names a challenge to reverse the entropy caused by the deregulation of finance and the unleashing of fossil fuels to support an economy of consumerism predicated on the perpetual exhaustion of desire, and the *proletarianization* of thought through its externalization in digital technologies (*For a New Critique* 33). The Great Acceleration of the postwar period replaces long-term circuits of intergenerational reproduction (e.g., memory and imagination) with short-term incentives that disrupt the attention spans required for reading, and that are necessary for the ability to "take care" of the technics which mediate individuals, society, and the earth. Stiegler proposes a task to transform the technics that shape social processes of individuation, which is another way of saying the way we read now (*Neganthropocene* 230). Stiegler was a student of Derrida's and Brennan a student of Said's, yet converging anthropogenic crises are bringing these traditions together. A recent collection edited by Rosi Braidotti and Paul Gilroy titled *Conflicting Humanities*, opens by reframing the posthuman through Said's humanism to address emerging axes of political antagonism and agency in the twenty-first century. These accounts of socio-ecological crises highlight the need to rearticulate the human within complex, multiple humanisms as they travel across interpretive contexts and shape reading practices.

The move from scale to antagonism ultimately tempers the idealisms of technological determinism and geological apocalypticism. Anthropocene scales too often lend themselves to "disaster theory," a mode of reading that discovers in every antagonism an *unreadability* that pushes toward aporias at the limits of thought. These readings range from ponderous, ontological introspections, to celebratory affirmations of "'left' extinctionism" (Cotter et al. 225). When scale reading transfers the figure of antagonism or contradiction from the social terrain to the realm of thought, the resulting struggle becomes not one of politics in the world, but rather the attempt to reconcile discontinuous temporal or spatial representations in the reader's mind. As they outpace apprehension, these "scale effects" generate conflicts in the realm of the possible rather than the actual. For ecocritic Timothy Clark, these "derangements of scale" produce a bewildering effect by multiplying overlapping temporal and spatial contexts. The "most prominent effect," Clark observes, is "an implosion of intellectual competences" ("Scale"). The text becomes a means to pose ever more scales that dramatize the limits of predictive knowledge, and the result is a declaration that its own failed thought is part of the disaster of the Anthropocene. The answer, as Clark suggests, "will not be a matter of inventing some new method of reading per se," but rather of challenging forms of "ideological containment" that constrain cultural work in university disciplines ("Scale"). Thus, close reading becomes a matter of putting disciplinary practices to additional ends, making narrative connections across antagonisms within the contexts of environment, work, and technology, that is, within *cultural* sites where the Anthropocene is reproduced rather than imaginatively overcome.

Freshwater socialism in the Pacific

In his final collection of writings, *Man Without a Country* (2005), Vonnegut juxtaposes a deeply pessimistic outlook on fossil fuels, climate change, and war, with a practical hopefulness. He characterizes this as "freshwater socialism," named after the politics of German immigrants in the Great Lakes states of the nineteenth and twentieth centuries. Originally a series of articles written at the nadir of the second Bush administration, the book ties together Vonnegut's concerns since World War II. His writing is defined by a struggle to maintain a critical solidarity with his fellow human beings as their collective actions race toward apocalyptic ends. His satires combine the playful, speculative textualities of fellow postmodern giants, yet their popular appeal lies in the mundane absurdities of daily life. *Galápagos* (1985) follows earlier critiques the displacement of human labor by automation in *Player Piano* (1952), and the atomic bomb (another possible Anthropocene marker) in *Cat's Cradle* (1963). *Galápagos* has been taken up by scholars to speculate about environmental scales beyond the human. However, Vonnegut's evolutionary frame draws into relief the antagonisms of daily life that structure the terrain on which multi-scale scenarios play out.

Galápagos glimpses a speculative evolutionary scenario driven by irrational events in which the scalar effects of actions run contrary to the intentions of characters. It describes an evolutionary bottleneck of a ship on the "Nature Cruise of the Century," the *Bahia de Darwin*. The collapse of financial markets into global war leaves global civilization vulnerable to a bacterial outbreak that eats the eggs of human females and results in the near extinction of humans. The novel presents a one-million year vision into a future populated by seal or dolphin-like post-humans, "fisher-folk," who are described as languid, semiaquatic beings, no longer troubled by large brains, thumbs, or the burden of language. Vonnegut's ghost narrator, the true subject of the story, questions what it means to move "beyond" the human, as he remains loyal in his study of humanity as it changes biocultural forms over the millennia. The ghost is Leon Trotsky Trout, son of Vonnegut's biographical caricature, Kilgore Trout. Leon is a U.S. soldier who refuses to continue fighting in Vietnam and moves to Sweden, where he later dies working on the construction site of the *Bahia de Darwin*. While his middle name is often omitted from commentary, this detail is introduced late in the novel, and returns the immense narrative scale to a singular, implicitly partisan frame within which the evolutionary future of capitalism, war, and empire, is presented. The cast of characters include wealthy Euro-Americans, a Japanese family whose daughter exhibits a furry mutation caused by the bombing of Hiroshima and Nagasaki, and Ecuadorian stowaways that include a group of young women who are indigenous and black descendants of slaves (described as the mothers of all future humans). Because of its evolutionary theme, *Galápagos* features a greater number of female characters than a typical Vonnegut novel. However, like the rest of the novel's outcomes, it is accidental. At every turn, Vonnegut positions the evolutionary logic of the narrative against expectations of harmony. What appears as a terrible act at one scale (e.g., after sex, the character Mary Hepburn manually inseminates other women against their knowledge or consent) becomes – in the evolutionary state of exception – the very thing that ensures future flourishing.

The Vonnegut papers at Indiana University include a draft "Preface to Galápagos" that provides additional clarity on the novel's politics of evolution. It describes his 1982 trip to the island with his wife, Jill Krementz, as she writes an ethnography of adoption. Vonnegut thanks her "for her honest responses to the unpleasantness of the islands and the moral blankness of Nature in the raw." Comparing the "quicksilver" adaptation of adopted children to their circumstances with the reptilian "clocks" of evolutionary time on the island, he is more impressed by the children than by evolutionary change that conserves form from one generation to the next. Mocking the notion that great scientific theories do more than reinforce the actions of the powerful, he speaks in the voice of evolution, commanding readers to "Prove by fucking or killing that you are Nature's favorite. Never mind mere human law." Returning to his own sarcastic voice, he warns: "This planet now faces extinction of all life forms by human wars and human overpopulation, which are returns to Nature supposedly – which are thrilling suspensions of mere human law. Stop thinking. Stop planning. Go crazy. Let Mother Nature decide" ("Preface"). As an indictment of the organized irrationalities of the militarized market society, the argument that planning is responsible for the Anthropocene makes a certain sense. However, Vonnegut likewise undermines the belief that Nature offers relief from the burdens of planning. For left libertarians, "going crazy" implies the restoration of a supposedly lost harmony between individuals and an uncoerced, productive nature. For right libertarians, "going crazy" implies the restoration of a supposedly natural order of property through the suspension of regulatory practices, viewed as necessary to establish a brutal new harmony. For those who think law and moral norms are perversions of natural order, Vonnegut's sarcastic wager to "let Mother Nature decide" suggests that the "thrilling" release of forces invites more risk than it is worth.

Reframed by the Anthropocene, *Galápagos* challenges the notion that Vonnegut retreated from political to "cultural matters" (Klinkowitz, *Vonnegut's America* 86). What is seen as an apolitical turn to nature and technology is now, at a later moment in the unfolding crisis, an outlook common to socialist ecology. "The idea . . . that nature's way is going to free us," as Andrew Ross puts it, "is a cruel inversion of the idea that liberation from nature is what frees us." Worse, it is perhaps "the very idea that helped get us into this mess in the first place." Vonnegut satirizes the dream of liberation from the weight of subjectivity which our "big brains" make possible. The only thing worse than having big brains – to remain with Vonnegut's lexicon – is not having one at all. "Neither nature nor economics are conscious agents," Ross continues, "our best future lies with an anti-Malthusian belief in the power of human needs and abilities to reorganize the social order of the natural and economic world" (272). Consistent with his freshwater socialism, Vonnegut's addresses from this period attest to human needs and abilities amidst the environmental dangers of technoscience and militarism.

In "Fates Worse than Death," a 1982 lecture at St. John the Divine Cathedral, Vonnegut advances a moral critique of the justifications for the nuclear arms race. Among these fates worse than death he lists are "life without petroleum" and subjugation by a foreign power ("Fates"). As the logic goes, these fates are unthinkable, worth dying to avoid – or, more accurately, worth nuking the planet to avoid – precisely because they save "us" from fates that might include slavery or serfdom. Affluent Euro-Americans, he argues, are oblivious to the fact that many of their fellow citizens experienced slavery and genocide, and yet have still found life preferable to the prospect of universal death. Linking the category of animality to the racialized violence that it so often underwrites, he observes that even blue-footed boobies in the Galápagos Islands prefer *animal* life to death. While visiting MIT in 1985, Vonnegut suggests hanging pictures of Mary Shelly "to remind students and faculty that humanity now cowers in muted dread, expecting to be killed sooner or later by Monsters of Frankenstein." "Such killing goes on right now," he continues, "in many other parts of the world, often with our sponsorship – hour after hour, day after day." Criticizing this complicity of the technical intelligentsia in the warfare state, Vonnegut proposes a "Hippocratic oath," choosing to meet people where they are instead of on a scale that estranges their relationships in the world: "The regimen I adopt shall be for the benefit of all life on this planet, according to my own ability and judgment, and not for its hurt or for any wrong. I will create no deadly substance or device, though it be asked of me, nor will I counsel such" ("Speech"). In the Anthropocene, the need to read, imagine, and dignify, other forms of life *as worthy of being lived* may serve to militate against the willingness to destroy others in order to preserve high-consumption lifestyles. Likewise, as long as reading remains driven by techno-scientific questions of scale rather than ethico-political questions of antagonism, it will mask a managerial ethos.

Antagonisms in *Galápagos*

Reading *Galápagos* in this context, we can better explore how the narrative stages antagonisms regarding nonhuman narrators, post-historical animality, and the transformation of reading by digital culture. As geologic scales reduce human history to a trace, they are joined by philosophies of inhuman perspectives. These philosophers ask whether or not "anti-human thought experiments finally offer a genuinely post-anthropological theory," or if they leave "some ghostly human subject still summoned in order to preside over its own . . . demise" (Bradley 20). Narrative theorists Pieter Vermeulen and Marco Caracciolo converge on similar questions regarding future, nonhuman readers. Vermeulen considers the epistemic and ethical value of nonhuman narrators as "future readers." Caracciolo argues that narrative does not necessarily reinforce anthropomorphic tendencies that close the reader off to ethical experience "beyond the human scale" (312).

Each raise ethical, epistemic, and ontological concerns, with the assumption that to think beyond the human is to be more ethical, as is the awareness of how our sensory organs problematically anthropomorphize this encounter *as* experience. Vermeulen posits two esthetic responses to these overlapping concerns. The first is a disjunctural sublime that attempts to radically unmoor anthropomorphic sense-making by way of nonhuman coordinates. In this account, literature and art should challenge the desire for narrative coherence because the planet, it follows, is radically *other* than our cognitive patterning. The second esthetic response reaffirms art's ability to narrate human agency in "safe-guarding" various forms of life on a "stubbornly nonhuman planet" (871). What is at stake in both is the possibility of an ethical commitment to a figure that is no longer recognizable, and unable to return our sympathetic imagination. Vonnegut takes the latter path. The movement of the novel's ghost narrator between intimate moments and evolutionary scales has produced divergent interpretations. Caracciolo maintains that "it is by leveraging the tension between these apparently irreconcilable positions that Vonnegut attains a truly posthuman – indeed, cosmic – vantage point" (311). Whereas *Galápagos* may go "beyond" human experience to the million-year view, it ultimately *returns* as it moves from the particular to the universal and back. This is a feature of consciousness, not a transcendence of it.

Reading *Galápagos* as an extinction narrative, Gerry Canavan finds a deep pessimism, which he links to contemporary "necrofuturological" postures and preoccupations with "melancholically sublime" death scaled up to the level of the species (136). Scale becomes a means for the novel to act out a "suicidal ideation" in which misanthropy is "optimistically" overcome by radically transforming the *anthropos* (140). Having witnessed the bombing of Dresden, the atomic age, and climate change, Vonnegut is no stranger to the impulse that Walter Benjamin observed, in which human "self-alienation has reached such a degree that it can experience its own destruction as an aesthetic experience" (242). Canavan laments that "no matter how estranged from our moment . . . the art object remains . . . a fundamentally human-centered one" (141). However, by affirming the anthropomorphic nature of narrative acts and future readers, Vonnegut's narrator maintains a fidelity to the human through a desire to communicate with the present. For example, the asterisks that mark the imminent death of characters, and their elimination from an evolutionary future, enable the anti-teleological evolutionary narrative to be shaped according to the unfolding linear plot of the cruise ship. *Galápagos* likewise closes on the narrator's decision to abandon the war and leave Vietnam for Sweden with the possibility of learning a new language rather than transcending humanity. Jerome Klinkowitz argues that the novel's "true optimism" resides in this act, "making it one of the most positive works in Vonnegut's canon" (*Vonnegut Effect*, 133). This alternate reading does not intuit the pseudo-utopian prospect of overcoming the human through extinction – which also eliminates the possibility of speculation, as *Galápagos* notes – but from the narrator's ability to act in an irreconcilable situation whose future could not be known in advance.

Connecting biology with these future nonhuman readers, Claire Colebrook warns that "the darting eye that stimulated the brain into becoming a reading and interpreting animal, may also be at the forefront of the human species' cognitive atrophy." In Colebrook's dark dialectic of the "self-extinguishing" brain, traits that enable survival (e.g., desire for fats and sugars) develop into their opposite, and are a dubious model for imagining "*what would be readable* for a world without readers" in the future "geological eye" of the Anthropocene (21). "The earth, after humans, will offer 'a reading' of a species' history," she argues, "just as we might say that *Robinson Crusoe* offers 'a reading' of race, empire and capitalism, even if neither Defoe nor his readers actually actualized the sense of the reading." This slippery usage of "reading" from a verb to a noun strives to eliminate the subject, overlooking the fact that it is *for us* – Defoe's future human, and ostensibly *left-wing* readers – that his text offers such a reading. This effort to redefine reading without even a body to do it describes a world without a subject. Vonnegut's ghost narrator is the more honest conceit in that it gives us a glimpse of post-humanity from a perspective that would be recognized as human by readers from ancient Greece to modern China. If extinction is a thought experiment, as these critics suggest, it is an experiment that affirms the obsolescence of culture in the face of geology and techno-capital, and therefore the human ability to transform these conditions, precisely because it misidentifies the human with the terraforming power of capital. Vermeulen illustrates this when he invokes Theodor Adorno to claim that the powerlessness of culture in the Anthropocene may be its source of insight (871). Yet when Adorno's argument about art's autonomy is understood in relation to his critique of the decline of the workers movement, this apparent valorization of powerlessness as an ethical virtue showcases how, as Brennan argues, theory misreads its own political retreat. The *displacement* of the human as an interpretive and willful historical subject falsely appears as a victory, at the very moment when the Anthropocene places the need for the collective making of history back on the table.

The vision of a post-historical animality in *Galápagos* appears only 4 years before Francis Fukuyama announces the end of history, a thesis inspired by Alexandre Kojève, who predicted that consumerism would return "Man" to animality through a collapse of the Subject that opposes itself to an Object (Kojève 161). Both predictions are disproven in the neoliberal era by the rise of speculation and scenario planning to shape the world as an object for market and military intervention. As a genre of scenario planning, science fiction moves within this speculative zeitgeist, yet *Galápagos* challenges popular assumptions about history and human identity. The belief, for instance, that social crises have their origin in biology is mocked to undermine the fantasy that animality holds an easy answer to socio-ecological conflict. Its self-reflexive narration carries warnings about big brains and the specter of decreasing brain size. The latter includes the loss of speculative capacities (both social and physical), concern for the future, and the sense of discordance between relationships as they are and as they might be. This antagonism turns inward, illustrated by character, Roy Hepburn: "I'll tell you what the human soul is . . . Animals don't have one. It's the part of you that knows when your brain isn't working right" (45). This potential to recognize an internal antagonism, either within the self or within a narrative, is the minimal condition for any critique of society. The novel's central conceit draws loosely on Vonnegut's "fates worse than death" thesis. It describes the easy life of the seal-like posthumans, yet nevertheless leans on the reader's recoil at this "soulless" resolution of history.

The potential fate of humanism as a practice is on display as Vonnegut's future humans have neither stories nor the means to share them. Entertainment comes mainly from unconscious bodily activities like hiccups and farting. There is no confusion about names or family heritage "since nobody has a name anymore – or a profession, or a life story to tell." They have "an odor" that is likewise incapable of lying, dissimulating, or fabricating (104). The most significant cultural difference is the loss of the biological ability to speculate, to "enjoy in their heads events which hadn't happened yet and which might never occur" (255). The capacity to imagine that things can be otherwise is not only of interest to science fiction as a genre, it is what enables the coordinated effort of narrative sense-making in the practical life of society needed to alter historical trajectories within the Anthropocene. Leon Trotsky Trout references a novel by his father titled *The Era of Hopeful Monsters*, in which the industrial destruction of the planet by humanoids is interrupted at the last minute by rapid evolutionary experimentation: "children with wings or antlers or fins, with a hundred eyes, with no eyes, with huge brains, with no brains," who eventually become "better planetary citizens than the humanoids" (86). Distinguishing between monstrous personalities and body types, *Galápagos* asks its readers to consider the upside of bodies evolved beyond the ability to oppress others. "How could you capture and torture someone with just your flippers and your mouth," Trout asks, or use weapons when "nobody today is nearly smart enough to make the sorts of weapons even the poorest nations had a million years ago" (156). He wonders the same about slavery, finance, and suicide. Vonnegut's evolutionary scale exaggerates the body as a cause of social exploitation, while confronting readers with the decline of narrative as a medium of antagonistic social exchange.

Galápagos raises the antagonism of the *medium* of reading in the digital age through the Mandarax, a solar-powered communication device that contains the entirety of human languages and literatures. For much of the novel, the Mandarax issues short, interruptive passages from great works of literature that serve as meta-commentary within the narration. Yet when the Mandarax is finally shown to intervene in the action of the plot, this handheld repository of human culture thwarts the characters' intentions. They are betrayed by this archive of culture, appropriately with a cryptic passage from a poem about betrayal. When they try to send an SOS signal, "mayday, mayday," the response they receive is a line from T.S. Eliot's poem, *Gerontion*: "In depraved May, dogwood and chestnut, flowering Judas,/To be eaten, to be divided, to be drunk/Among whispers" (271). Readers commonly seize on the allegory of the Mandarax as the "Apple of Knowledge" cast into the sea by Captain von Kleist as the New Adam. However, the captain's act is not motivated by the threat of too much knowledge but by anger at the "uselessness of all its knowledge" (63). The technical ability to accumulate and store information transforms the relationship of the reader to the text. It collapses the distinction between reading for information (search or recall) and reading for cultural knowledge. This diminishing utility is, as Stiegler argues, also a form of proletarianization, as it separates people from *savoir faire* (practical knowledge) and *savoir vivre* (knowledge of living well) (*For a New Critique*, 16). Vonnegut's apparent lack of faith in *nonhuman* processes, whether natural or technological, suggests that cultural knowledge is necessary to understand how objects and practices may be threatened by their promised means of archival preservation.

The materiality of the digital in the Anthropocene redraws long-existing distinctions between form and meaning that must now include the prosthetic supports for human culture. In his exchanges with Katherine Hayles on modes of attentive reading, Stiegler argues that "text, as the principal support for deep attention, has become a new kind of contextuality" (2010, 83). In other words, by attending to the cognitive practices of reading, we are increasingly drawn into the ensemble of relations that make it possible. This means including the development of human senses in relation to the history of technics, and to earthly contexts of material and cultural production, in an expanded philology. This supports Brennan's argument for a philology that grounds meaning in the lived history of peoples rather than in technological and deterministic formalisms. Close reading traces the antagonistic attachments of culture to the material histories of the Anthropocene, which began long before it was named.

Conclusion

Reading Anthropocene texts through projections of scale can produce compelling speculations about the possibilities for human and nonhuman futures. However, the shift to antagonism enables critics to read the social terrain in which the struggle for the future is waged. Articulation makes this move from scale to antagonism possible, not through a narrowing of vision or ethical concern, but by identifying contiguities with the past as well as futures in the making. Moreover, it tempers the bipolar tendency to either celebrate or exterminate the human *in theory* as a response to frustrations *in practice*. Brennan reminds us that "the everyday life of those living in U.S. and Western capitalist societies is not at all characterized by an exaggeration of human capacities" (232). By situating authors, narratives, and readers, within the discourses and structures of feeling of a historical moment, this approach illustrates how a novel might speak to its readers in other times and places. Despite the evolutionary scale of *Galápagos*, its criticism of U.S. finance and imperialism (the lingering traumas of Vietnam) aims the novel squarely at the heart of the political community. It raises the necessity and limits of individual action in the context of an apparent feedback loop of war and ecological degradation. It reminds us that Darwin's own anti-teleological outlook was shared by Friedrich Engels, who wrote that "*History* does *nothing* . . . it 'wages *no* battles.' It is *man* [sic], real, living man, who does all that, who possesses and fights; 'history' is not, as it were, a person apart, using man to achieve *its own aims*; history is *nothing but* the activity of man pursuing his aims." The irony is that while present actions bear such an outsized effect on future ecological history, it remains a future written behind our backs.

Since *Galápagos* was published, over half the carbon ever released has entered the atmosphere. If Fredric Jameson's observation that contemporary culture makes it easier to imagine the end of planetary life than the end of capitalism is correct, then, surely, interpretive practices that erase the human are an accelerated route to this same impasse. Against this double erasure, close reading in the Anthropocene articulates the human in its many-sided contexts, highlighting the antagonisms that bifurcate the *anthropos*, and refusing scales that pretend to know the outcome of struggles before they have even been fought.

Notes

1 Stuart Hall (1983/2016) develops the practice of articulation from Antonio Gramsci in response to debates between E.P. Thompson and Louis Althusser over antihumanism (167); see also Jennifer Daryl Slack in *Stuart Hall: Critical Dialogues in Cultural Studies* (1996), Grossberg's *Cultural Studies in the Future Tense* (2010), and John Clark (2015).

2 Brennan considers the non-Western sources of humanism that informed the anti-colonial precursors to Marxist thought, against the Cartesian, Spinozist, and Nietzschean precursors of posthumanism; see also "Philology" (2014).

Works cited

Agamben, Giorgio. *The Open: Man and Animal*, translated by Kevin Attell, Stanford University Press, 2004.

Benjamin, Walter. "The Work of Art in the Age of Mechanical Reproduction." *Illuminations*, edited by Hannah Arendt, Shocken, 1969, pp. 217–51.

Boltanski, Luc and Eve Chiapello. *The New Spirit of Capitalism*, translated by Gregory Elliott, Verso, 2005.

Bonneuil, Cristoph, and Jean-Baptiste Fressoz. *The Shock of the Anthropocene*. Verso, 2016.

Bradley, Arthur. *Originary Technicity: The Theory of Technology from Marx to Derrida*. Palgrave Macmillan, 2011.

Braidotti, Rosi and Paul Gilroy, editors. *Conflicting Humanities*. London: Bloomsbury, 2016.

Brennan, Timothy. *Borrowed Light: Vico, Hegel, and the Colonies*. Stanford University Press, 2014.

———. "Philology." *ACLA State of the Discipline Report*, Mar. 2014. Accessed 3 February 2020. stateofthediscipline.acla.org/entry/philology

Canavan, Gerry. "After Humanity: Science Fiction after Extinction in Kurt Vonnegut and Clifford D. Simak," *Paradoxa*, vol. 28, 2016, pp. 135–56.

Caracciolo, Marco. "Posthuman Narration as a Test Bed for Experientiality: The Case of Kurt Vonnegut's *Galápagos*." *Partial Answers*, vol. 16, no. 2, 2018, pp. 303–14.

Cohen, Tom, et al. *Twilight of the Anthropocene Idols*. Open Humanities Press, 2016.

Colebrook, Claire. *Death of the Posthuman: Essays on Extinction, Vol. 1*. Open Humanities Press, 2014.
Cotter, Jennifer, et al. "The 'Event-al' Logic of Disaster and 'Left' Extinctionism." *Human, All Too (Post)Human: The Humanities after Humanism*, edited by Cotter et al., Lexington, 2016, pp. 223–35.
Clark, John. "Stuart Hall and the Theory and Practice of Articulation." *Discourse*, vol. 36, no. 2, 2015, pp. 275–86.
Clark, Timothy. "Scale." *Telemorphosis: Theory in the Era of Climate Change*, edited by Tom Cohen, Open Humanities Press, 2012.
During, Simon. *Exit Capitalism: Literary Culture, Theory, and Post-Secular Modernity*. Routledge, 2010.
Engels, Friedrich. *The Holy Family*. 1844. Accessed 1 March 2020, marxists.org/archive/marx/works/1845/holy-family/ch06_2.htm
Greif, Mark. *The Age of the Crisis of Man: Thought and Fiction in America, 1933–1973*. Princeton University Press, 2017.
Grossberg, Lawrence. *Cultural Studies in the Future Tense*. Duke University Press, 2010.
———. *We Gotta Get Out of this Place: Popular Conservatism and Postmodern Culture*. Routledge, 1992.
Hall, Stuart. *Cultural Studies 1983: A Theoretical History*, edited by Jennifer Daryl Slack and Lawrence Grossberg, Duke University Press, 2016.
Klinkowitz, Jerome. *Kurt Vonnegut's America*. University of South Carolina, 2009.
———. *The Vonnegut Effect*. University of South Carolina, 2004.
Kojève, Alexandre. *Introduction to the Reading of Hegel*, edited by Allan Bloom, translated by James Nichols Jr., Cornell University Press, 1958.
Leitch, Vincent. *American Literary Criticism since the 1930s*, 2nd edition. Routledge, 2009.
Lewis, Simon, and Mark Maslin. "Defining the Anthropocene." *Nature*, vol. 519, March 2015, pp. 171–80.
Menely, Tobias, and Jesse Oak Taylor, editors. *Anthropocene Reading: Literary History in Geologic Times*. Pennsylvania State University, 2017.
Nealon, Jeffrey. *Post-Postmodernism: Or, The Cultural Logic of Just-In-Time Capitalism*. Stanford University Press, 2012.
Ross, Andrew. *The Chicago Gangster Theory of Life: Nature's Debt to Society*. Verso, 1994.
Said, Edward. "The Return to Philology." *The Selected Works of Edward Said, 1966–2006*, edited by Bayoumi and Rubin, Vintage, 2019, pp. 529–51.
Schryer, Stephen. "Fantasies of the New Class: The New Criticism, Harvard Sociology, and the Idea of the University." *PMLA*, vol. 122, no. 3, 2007, pp. 663–78.
Slack, Jennifer Daryl. "The Theory and Method of Articulation in Cultural Studies." *Stuart Hall: Critical Dialogues in Cultural Studies*, edited by David Morely and Kuan-Tsing Chen, Routledge, 1996, pp. 112–27.
Stiegler, Bernard. *For a New Critique of Political Economy*, translated by Daniel Ross, Polity, 2010.
———. *The Neganthropocene*, translated by Daniel Ross, Open Humanities Press, 2018.
———. *Taking Care of Youth and the Generations*, translated by Stephen Barker, Stanford University Press, 2010.
Swanson, Heather Anne, et al. "Less Than One but More Than Many: Anthropocene as Science Fiction and Scholarship-in-the-Making." *Environment and Society*, vol. 6, no. 1, 2015, pp. 149–66.
Tumino, Stephen. *Cultural Theory after the Contemporary*. New York: Palgrave, 2011.
Toivanen, Tero et al. "The Many Anthropocenes: A Transdisciplinary Challenge for Research." *The Anthropocene Review*, vol. 4, no. 3, 2017, pp. 183–98.

Vermeulen, Pieter. "Future Readers: Narrating the Human in the Anthropocene." *Textual Practice*, vol. 31, no. 5, 2017, pp. 867–85.
Vonnegut, Kurt. "Fates Worse than Death." *The Spokesman*. Accessed 12 December 2019. www.spokesmanbooks.com/Spokesman/PDF/92Vonnegut.pdf
———. *Galápagos*. Dial Press, 1985.
———. *Man without a Country*. Seven Stories Press, 2005.
———. "Preface to Galápagos." *Vonnegut mss., 1941–2007*. LMC 1860, Indiana University Archives, Bloomington.
———. "Speech at M.I.T." *Vonnegut mss., 1941–2007*. LMC 1860, Indiana University Archives, Bloomington.
Whyte, Kyle. "Indigenous Climate Change Studies: Indigenizing Futures, Decolonizing the Anthropocene." *English Language Notes*, vol. 55, no. 1–2, 2017, pp. 153–62.
Žižek, Slavoj. "Holding the Place." *Contingency, Hegemony, Universality: Contemporary Dialogues on the Left*, Verso, 2000, pp. 308–29.

INDEX

Note: Page numbers in *italics* indicate a figure on the corresponding page.

"aboutness" 33
activist-oriented speculative readings 177
actor-network theory 86, 89n17
adequacy, state of 110–11
"administered world" 34
Adorno, Theodor 34–5, 69, 184
affective labor 54
affirmative critique 134, 141
Africa: national security formations 97; Nigeria 9, 95–100; World War II era 34, 94; *see also Burma Boy* (Bandele)
Agamben, Giorgio 120, 123, 175
agency: autonomy 26, 82, 84–5; distributed 83; multispecies 7–8, 60; of naming 96–7; of narrative 95; nonhuman 62–3, 72n27, 79, 90n19; of student 81
age of the world target 9, 116, 119, 122, 128–9
Alaimo, Stacy 62
Allenby, Braden 171
Altars for Peace (Nakashima) 9, 116, 120–30; at Cathedral of St. John the Divine *121*, 126, *126*; conception (1969–1983) 122–5; development (1984–1989) 125; as diplomacy 127–8; final (1990–present) 128–9; Foundation 126; to Hiroshima or Nagasaki 126, 129; Moscow 126–8; walnut tree used in 120, *121*, 123–5, *125*, *126*
American Terror (Hurh) 67–8
analogy 31–4, 48, 63, 69; as figure of occupation 63; in Stevens's writing 107–8, 110
Angus, Ian 113n1
animality 11, 154, 182; post-historical 175, 183–5
Anker, Elizabeth 133
antagonism: between capitalism and the earth 11, 175; and double erasure 11, 175–9, 187; in *Galápagos* 182–6; move from scale to 175, 179, 186
Anthropocene: Atlantic Slavery as first step in 93; closure, impossibility of 94; as cloven concept 10, 131–2; dating onset of 3, 61; disorienting framing of 77; eclipsed sun image in conjunction with *152*, 152–3; extinction overlooked 2, 62; and Ford's paintings 47–8; as historical category 160; humanist responses to 166; humanity not at center of 132; limits of as term 15; as literary object 79; made in image of "Man" 9, 60; as "millennial concept" 177; nonhuman, recontextualization in 7, 9; planning as cause of 181; poetics of 76, 102; provincialization of 7, 48; reading in/of/for 7, 48; reproduction of in cultural sites 179; as subject of close reading 3; as term 1–2, 15; as undifferentiated social stratification 12, 62, 73n38, 139–40; as unreadable

134, 138, 179; as unthinkable 5, 70, 85, 90n19, 182; as world-maker/destroyer of worlds 98
Anthropocene disorder 61
"anthropoceneologists" 177
Anthropocene Reading (Menely and Taylor) 77, 87n3
Anthropocene Working Group (AWG) 2–3
Anthropos: as buried 10, 175; and double erasure 11, 175–9, 187; drive for mastery 110; literature as possibility for understanding 5; originary technicity of 178–9; social difference elided under 11, 175; "we" of 2, 62; *see also* human
Apocalypse Now (film) 153
Arcades Project, The (Benjamin) 147
archeology, narrative 94–100
Arctic Dreams (Lopez) 26
Aristotle 18, 69
articulation 176, 179, 186–7
Arts of Living on a Damaged Planet, The (ed. Tsing) *152*, 152–3
artwork: aura of original 145–6; habitat, careful attention to 50, 51; as human-centered 183; materiality of 129; nonrealistic 160; reading 48–9; scale of 161; *see also* Ford, Walton; "Funk Island" (Ford); "Witch of St. Kilda, The" (Ford)
astrology 156
Atlantic Slavery 93
atomic imaginary 119–20
attachment 83–4, 86, 86–7n1
attention 4, 7–9, 11, 165, 172; anti-anthropocentric 148; attunement 26–8; as beginning of ethics 110; in Bishop's work 110–13, 114n6; categories of 47; close reading associated with 81; commons of 84; deep 78–83, 87n8, 186; extended 4, 7, 47–9; flânerie as 147–50; and flow 27–8; and Great Acceleration 179; to habitat, in Ford's work 50–1; to lived experience 83–4; New Criticism's focus on 103–4; reconfigured 150; "search image" in brain 150; in Stevens's work 106–7; to text 22–3, 133; unity of dual focus 27
"At the round earth's imagin'd corners" (Donne) 31
Attridge, Derek 36, 41, 42
attunement 26–8
Auden, W. H. 4–5
Audubon, John James 47
auk *see* great auk extinction
aura, of original artworks 145–6
Auroville, India 128
autonomy 26, 82, 84–5

bad faith 8, 62
Badiou, Alain 61
Balaam, Peter 9
balance 61, 67–9, 72n33
Bandele, Biyi 8–9, 93–101; anti-neocolonialist interpretation of 95; as diasporic writer 94; *The Man Who Came in from the Back of Beyond* 95; *The Sympathetic Undertaker and Other Dreams* 95; *see also Burma Boy* (Bandele)
Barthes, Roland 4
Bartkevicius, Jocelyn 151–3, 154
"Bartleby Industry" 64, 68
"Bartleby Industry and Bartleby's Idleness, The" (Knighton) 68–9
"Bartleby, the Scrivener: A Story of Wall Street" (Melville) 7–8, 60–75; Anthropocene anticipated in 63; balance in 61, 67–9, 72n33; becoming rock in 62, 66–8; duality of 63–4, 68–9; geologic figures in 63, 66–8, 72n26; inhuman agency in 62–3, 72n27, 79, 90n19; interconnectedness 63–4, 66; island metaphor in 63–6, 69; narrator as Bartleby's double 67; pathologization of Bartleby 62–3; substance of 69–70; Turkey and Nippers 68; *see also* Melville, Herman
Baucom, Ian 94
Beck, Ulrich 137, 177
Being and Time (Heidegger) 163
Benjamin, Walter 145–8, 183; *The Arcades Project* 147; "Doctrine of the Similar" 156–7; "The Work of Art in the Age of Its Technological Reproducibility" 145
Berardi, Franco "Bifo" 177
Berlin Wall, fall of 116, 128
Bernstein, Leonard 126
Best, Stephen 88n14, 146
beyond the human 5, 25, 70, 147, 151–2, 169; in *Galápagos* 177, 179, 180, 182–3
Bhabha, Homi 4
Bialostosky, Don 81
Billion Black Anthropocenes or None, A (Yusoff) 2, 62, 93–4, 139
Binding, Paul 163, 172n4
Birds of America, The (Audubon) 47
"Birds Returned, The" (Smith) 7, 37–9, 40
Bishop, Elizabeth 9, 102–15; attention in work of 110–13, 114n6; "The Bight" 107; expression and form integrated

in 108; "The Fish" 111–12; "Florida" 107; groundedness of poetry 107–8; "In the Waiting Room" 110; "Jerónimo's House" 110–11; landscape as existential anchor in 110–11; "Little Exercise" 111; *North & South* 106–7; "One Art" 113; "Sandpiper" 110; "Seascape" 110; "The Unbeliever" 111
bitterbrush 118–19
black humanity: conscription of 94, 97; as inhuman matter 93, 139; *see also* Africa; Burma Boy (Bandele)
Blau, Sheridan 81
Bleak House (Dickens) 68, 72n34
Bonneuil, Cristoph 177
Borrowed Light: Vico, Hegel, and the Colonies (Brennan) 178
Bradley, Arthur 182
Braidotti, Rosi 10, 134–5, 141
Brassier, Ray 172
Braudel, Fernand 164
Brecht, Bertolt 34
Brennan, Timothy 178, 186
British Empire 30, 34, 94, 96
Brodkey, Linda 85
Brooks, Cleanth 4, 21–2, 103, 146, 165
Buddhist doctrines 168–9
Bunyan, John 111
Burma Boy (Bandele) 8–9, 93–101; child soldiers in 95–100; and colonial extractivism 93–4, 96; naming in 96–7; petrol discourse in 98–9; Yoruba metaphor of hurriedly buried corpse 9, 95, 100

Calvino, Italo 1, 3, 12
Cambridge Companion to Literature and the Anthropocene (ed. Parham) 5
Cambridge study of response to poetry 30–1
Canavan, Gerry 183
"Canonization, The" (Donne) 103–4
Can Poetry Save the Earth? A Field Guide to Nature Poems (Felstiner) 36, 103
capitalism 72n34; antagonism between earth and 11, 175; anti-instrumentalist art in response to 34–5; bad faith of 8; blame of entire species for white male actions 2, 62, 94; end of planetary life easier to imagine than end of 11, 187; global 118, 177–8; ideals of 27; naturalization of 177; neoliberal, predatory supremacy of 61–2; patriarchal 61; "totalizing" system of 34
Caracciolo, Marco 94, 98, 173n7, 182, 183
Carson, Anne 10, 154–6
Castronovo, Russ 63, 69, 71n13
categorization 66, 93, 139, 141
Cathedral of St. John the Divine *121*, 126, *126*
Chakrabarty, Dipesh 157, 172n1, 173n6
Charter of the Forest 124–5
children, burden of disease borne by 14n14, 70n2
child soldiers 8; African fiction boom 95–6; double vulnerability of 99; in postcolonial wars 100; in World War II 95–100
Chow, Rey 9, 116, 119, 120, 123
chthonic powers 141
Chu, Seo-Young 160, 161
Chua, Rina Garcia 7
civic action 9
Clare, John 165
Clark, Nigel 61, 62
Clark, Timothy 10, 41, 61–2, 79, 87–8n9, 87n5, 134, 179
climate change 37, 79–80, 87n9, 95, 157, 180, 183; children bear burden of disease 14n14, 70n2; "critical climate change" series 177; poesis of 6–7, 20; poetics of 40–1; rhetoric of 16–17
climate change denial 6, 17–18
climate change writing 6, 41
"climate generation" 80
climate wisdom 16
"close," as term 4
closeness reading 146–8, 153
close reading: allowing the text to be close to us 23–4; as assemblage of practices 176; attachment 83–4, 86, 86–7n1; broadened 8, 9, 94; Cambridge study of response to poetry 30–1; as collective practice 8, 77, 78, 80, 82–6; as critique 131–2; definitions 80–1; disengaging from form 78; and distant reading 4, 11–12, 132, 146; "doing" a close reading 81; as eco-pedagogy 40–3; as embodied practice 42–3, 84; engagement with text itself 21–4, 137; for estrangement 35–7; as event 83, 85; framework of assumptions about 76, 89n17; groundedness of 107–8; and lived experience of reader 83–4; as meaning-making 81, 83, 88n13; as mode of production 176–7; mystery, openness to 28–9; of nature 104–7, 151; paradoxes of 81, 84–5, 144, 145–6; as personal 23–4; phenomenon-focused

approaches 78; pragmatist 6, 16, 22–4; as relational activity 8, 19–20; at sentence level 23; as term 3–4, 144; transactional approach 82–4, 86; of trees 119–21; without critique 132; *see also* New Criticism; pragmatist close reading; reading
Close Reading (Lentricchia and DuBois) 3–4
Cloud Atlas of 2004 (Mitchell) 168
cloven concepts 10, 131–2
cognitive dissonance 159, 165–6, 171
Cohen, Jeffrey 78
Cohen, Tom 177
Colebrook, Claire 166, 184
Coleridge, Samuel Taylor 71n24, 103; "Frost at Midnight" 104; "Kubla Khan" 106
collective, concept of exiled 85
collective reading 8, 77, 78, 80, 82–6; reading groups 90n20; transactional approach 82–4, 86
colonialism 2, 47, 54, 62; extractivism of 93–4, 96, 139; imbrication of humans and nature 94; naturalization of 177; World War II era 34, 94–100
commons 10, 63, 84–5
Communist Manifesto (Marx) 62
Conflicting Humanities (ed. Braidotti and Gilroy) 179
conjunctive reading 11, 145, 151–7
consciousness 26–7, 66, 110–11, 162, 183; embodied 5; evolution of 166; global 64; transformational 8–9
consent 123–5, 129
correspondence 102–3
Counter-Desecrations: A Glossary for Writing Within the Anthropocene 19
Covid-19 pandemic 12
Crane, Louise 104
Crawford, Matthew 84–5, 89n18
creativity 5, 10, 29, 42, 84, 144–5
Critchley, Simon 171–2
critical climate change 177
critical posthumanities 134
critical practice 12, 49, 79, 159
critical race theory 139
critique 5; affirmative 134, 141; close reading as 131–2; critique of 10, 133; denaturalization of 84, 133; as form of cultural work 77–8; scale critique 11, 159, 161
cross-species conflict 47
cryptography 145–6
Culler, Jonathan 72–3n36, 76
cultural criticism 77, 79, 133–4
cultural knowledge 186
culture of literacy 85–6
current-traditional rhetoric 82

damaged planet 16, *152*, 152–3
Darwin, Charles 107, 187
Dasein 163
data-driven analysis 145–6, 153, 173n5
Davies, Jeremy 132, 139, 140, 141
decoding 20–2
deep attention 78–83, 87n8, 186
deep reading 8–9, 10, 131–43
deep thinking 36, 139
deep time 51, 64, 78, 86n1, 135–9
defamiliarization (*ostranenie*) 34, 36, 42
"Defense of Fiction" (Smith) 24–5
Defoe, Daniel 110, 184
deforestation 3, 150
Deleuze, Gilles 67
DeLoughrey, Elizabeth 7, 48, 94
de Man, Paul 165
democratic states 33, 120
denaturalization of critique 84, 133
depoliticization 63, 176, 178
derangements of scale 11, 79, 159–74, 179
"Derangements of Scale" (Clark) 79
Derrida, Jacques 70, 173n10, 179
de Sade, Marquis 49
despair, and acquisition 18
Dickinson, Emily 162
digging, as primal activity 139
DiLeonardi, Sean Michael 145–6
Dillard, Annie 10, 148, 150–1; *Pilgrim at Tinker Creek* 150; "Total Eclipse" 10, 151, 153–4
disaster theory 179
dissent, acts of 123–4, 129
distance 132, 144–5
distant reading 4, 11–12, 61, 89n17, 173n5; and eclipse writing 146–8, 151, 153; and "The Passing Strange" 159, 164, 173n5
Donne, John 31, 103–4
double erasure 11, 175–9, 187
Duarte-Gray, Isabel 9–10
DuBois, Andrew 3–4, 104
Duineser Elegien (Rilke) 162
During, Simon 176

Eagleton, Terry 34–5
Earth: as damaged planet 16, *152*, 152–3; as ground of ethics and politics 61; human ancestral relationships to landscape 138–9; humanity, relation

to 61, 62, 66, 67, 71n24; humans embedded on 5; as inhospitable 67, 70; no longer a backdrop 102; *see also* world
"Earth system humanism" 79, 87n8
Easterlin, Nancy 35
eclipse *see* total solar eclipse
Ecocollapse Fiction and Cultures of Human Extinction (McFarland) 6
ecocriticism 4, 9, 48, 141n1
Ecocriticism on the Edge (Clark) 61–2, 134
ecological thinking 9, 27
economic growth, argument against 18
ecopoetics 7, 114n3; eco-pedagogy 40–3; myth of motivated ecopoetic form 31–4; "negentropic" qualities attributed to poems 31–2; *see also* poetry
Ecopoetics (journal) 42
Ecopoetics: The Language of Nature, the Nature of Language (Knickerbocker) 36
Ecopoetry Anthology, The (ed. Fisher-Wirth and Street) 37
ecostudies 7, 48
ecosystems 9, 25, 81; poetic structure, analogy with 32
Eisenhower, Dwight 127
Electoral College 144
elegy 38, 53–4
embodiment 33, 84, 98, 112, 139; in "Bartleby, the Scrivener" 63–5, 70; body as cause of social exploitation 185; and close reading 42–3, 84; of consciousness 5; digital disembodiment 11, 175; geology of 70; honoring of 112; nonhuman 25, 156; of perception 147; scale of 160
emergency, practice of 42
Emerson, Ralph Waldo 16, 18, 21, 109
"Encantadas, or Enchanted Isles, The" (Melville) 60, 64, 71n6
"energy," discourse of 31–2
Engels, Friedrich 187
environment: poem as 33, 35; as term 33
environmental humanities 4, 93; proposals for expanding humanist practices in 77–8; scale critique 11, 159, 161
"environment poems" 36
epistemologies: indigenous 25–6; radical 10, 134–5, 140, 141
esthetics 21, 24–5, 72n37, 83, 87n7, 148, 176, 183; Marxist 34–5; in Nakashima's work 116–20, 122
estrangement 35–7, 40, 162
eternal return 1–2
ethics 161; attention as beginning of 110; Earth as ground of 61; evolution of 171–2; of inhabitation 8, 61–3, 65, 69–70; intergenerational 168
event, close reading as 83, 85
evolution 156; in *Galápagos* 11, 175, 180–6; rupture in 46
evolutionary narrative 11, 175, 183
extinction 2, 13n12; defining 46–7; different kinds of 46–7; "each time unique" stories 47, 54; material-discursive archives of 77; multispecies 5; overlooked 2, 62; resistance to 6; self-extinction 11, 166, 183; Sixth Mass Extinction 4, 13n8, 60; as thought experiment 183; *see also* great auk extinction
extinction archive 49, 52, 54, 55n18
extinction narratives 183
extinction studies 7, 46–7, 52–4
extractivism 8–9, 10; and children 97–8; colonial 93–4, 96, 139; covering of remains 94–5; Macfarlane's critique of 137; and "White Geology" 8, 93–5
eye, geological 184

fabliau 108–9
fabulation, speculative 177
Farrier, David 76, 161
Fate of the Earth, The (Schell) 122–3, 127
"faulty thinking" 18
Felski, Rita 84, 86, 88n12, 89n17, 133
Felstiner, John 36, 103
feminist theory 137
figurative language 37
Fisher-Wirth, Ann 37
flâneur 147–50, 156
flat perspectives 135
Fletcher, Angus 32–4, 35
Florida *see* Key West, Florida
flow 27–8
foraging 150
Fordism 176–7
Ford, Walton 7, 47–59; *Birds of America* in paintings of 47; "Funk Island" 7, 47–54; grievable subjects in works of 52; lastness in works of 52–3; limited scholarly study of 47–8; provincialization of Anthropocene 7, 48; story templates in works of 47, 54; "The Witch of St. Kilda" 47, 48, 51–4
form: disengaging close reading from 78; and meaning 36–9; Russian Formalism 35–6; *see also* motivated form, myth of
Frankfurt school 34
free verse 34–5, 39–40
Fressoz, Jean-Baptiste 177

Friedman, William F. 145
Fukuyama, Francis 184
Funk Island *see* great auk extinction
"Funk Island" (Ford) 7, 47–54; description of 49–52; story template of 54; written elements 54
furniture design 117
futurability 177

Gadda, Carlo Emilio 12
Gagliano, Monica 18, 25
Galapagos 64
Galápagos (Vonnegut) 175–89; antagonisms in 182–6; as evolutionary narrative 11, 175, 180–6; as extinction narrative 183; "freshwater socialism" in 180–2; ghost narrator 180, 183–4; optimism in 180, 183; pessimism in 183; speculation's role in 177, 180, 183–5
Galison, Peter 119–20
Gallop, Jane 104
Garrard, Greg 7, 42
Gaskill, Nicholas 78, 87n7
Gathering Moss (Kimmerer) 148
Gelpi, Albert 102–3, 107, 109, 113n2
Geneva Summit (1985) 127
genres: climate change writing 6, 41; nature writing 136, 138, 145, 148, 150–1; new nature writing 141n1; *see also* novel; poetry
geologic force, human as 63, 73n38, 138–9
geology 8–10; afterlife of 9, 94–5; as corporeal 70, 72n37; as extractive mechanism 93–4; insurgent 95; stratigraphic signals 2–3, 48
George VI 96
George Nakashima: Full Circle (Ostergard) 117
German immigrants, Great Lakes region 180
Ghosh, Amitov 85, 89–90n18
Gilroy, Paul 179
global consciousness 64
globalization 118, 177–8
golden spikes 3, 7, 54
Goldman, Irene 128
good conscience, complacency of 11
Gorbachev, Mikhail 127
Graham, Stephen 135
grand narrative 104, 139, 177
Great Acceleration 3, 179
great auk (*Pinguinus impennis*): Funk Island colony 45, 47, 53, 54n1; harvesting of 45–6; *IUCN Red List*'s hybrid extinction date 45–6
great auk extinction 7, 45–59; violence of 46–7, 50, 52–4; weight of 53; *see also* Ford, Walton; "Funk Island" (Ford)
Great Derangement, The: Climate Change and the Unthinkable (Ghosh) 85
greenwashing 176
Griffiths, Matthew 40–1
Grossberg, Lawrence 176
GSSP (Global boundary Stratotype Section and Point) 2–3
Guillory, John 78, 79–80, 81, 88n10
Gullestad, Anders 67
Gunster, Shane 167
Guyer, Sara 165, 173n5

Hague Appeal for Peace 128
Hall, Stuart 96, 176
Hamilton, Clive 102
happy endings 6
Haraway, Donna 15, 141, 177
Hardwick, Elizabeth 63
Harvey, David 132
Hass, Robert 40
haunting (ghostliness) 62, 64–5, 69–70, 136
Hawthorne, Nathaniel 19–20, 23
Hayles, Katherine N. 78–9, 87n8, 186
heavenly bodies 10–11
Hegel, Georg Wilhelm Friedrich 164
Hegglund, Jon 166
Heidegger, Martin 163
Heise, Ursula 47
Hemingway, Ernest 112
hermeneutics 76, 132
History of Newfoundland (Thomas) 46, 50
Hitler, Adolf 120
Hobbes, Thomas 123
Holocene 3, 9, 48, 102, 112
Homogenocene 3, 13n8
House of the Seven Gables (Hawthorne) 19–20
Huggan, Graham 10
human: blame of entire species for white male capitalist actions 62, 94; as cause and site of suffering 11; displacement of 184; dissolution of 177; inhuman interdependence with 63; undoing of as necessary 134; *see also Anthropos*
human exceptionalism 5, 47
humanism 66, 94, 178–9, 185; bad faith of 8, 62; Earth system 79, 87n8; potential fate of on display 185; universality of 63
humanity: African 93–4, 97; Bartleby as figure for 63, 67; common 67, 93;

Earth as constitutive of 61, 138–9; emergence of from underground 138–9; as geologic force 63, 138–9; not at center of Anthropocene 132; stratified 12, 73n38, 138–9
"human resources" 94
Hunt, Will 10, 134, 138–41
Hurh, Paul 67–8
hyper attention 87n8

Ice Age, last 104
identity 21–2, 28, 41, 89n19, 90n20, 184
ideological containment 179
imagination 24, 36, 83, 89n19, 102–3, 113n2, 136–7; poetic 106, 111
imperialist posthumanism 178
Index of Coincidence and Its Applications in Cryptanalysis, The (Friedman) 145
indigenous peoples 3, 15–17, 105, 180; ecological practices of 25–7, 148; Wajarri people (Western Australia) 140
inhabitation 8, 64, 71n6; of ecologies of attention 84; ethics of 8, 61–3, 65, 69–70; of words and worlds 6–7, 16, 21, 29
inheritance 137–8
inhuman/nonhuman 61–2; agency of 62–3, 72n27, 79, 90n19; Earth as 61; and inhabitation 8, 62; interdependence with human 63; materiality of 62, 70, 93–4; nonhuman narrative 77, 182–3; philosophies of 182; as race 93–4; recontextualization of Anthropocene in 7, 9; reduced to object 26; within human 68, 70
"In Memory of W. B. Yeats" (Auden) 4–5
instrumentalism 7, 37, 39, 41, 46, 150–1
"Intentional Fallacy, The" (Wimsatt) 145
interconnectedness 63–4, 66
interdisciplinary/transdisciplinary approaches 21, 34, 48–9, 177
intergenerational reproduction 179
intersubjectivity 7, 16, 25
intertextuality 53
island metaphor 63–7, 69–70
IUCN Red List of Threatened Species 45, 46

James, William 6, 16, 19, 22
Jameson, Frederic 11, 187
Jamieson, Dale 102
Jeffers, Robinson 6
jellyfish (*Aurelia aurita*) 39–40
Joyce, James 164
Justine, or Good Conduct Well Chastised (Sade) 49

Kafka, Franz 70
Kapitsa, Sergei 116, 128
Karma 168
Key West, Florida 9, 102–15; as art colony 104; as *Cayo Hueso* (bone island) 104; sameness of 104–5; *see also* Bishop, Elizabeth; Stevens, Wallace
Kimmerer, Robin Wall 25–6, 27, 148
King's African Rifles 94
Klinkowitz, Jerome 183
Knickerbocker, Scott 36, 42–3
Knighton, Andrew 68–9
Knorr, Don 117
knowledge, as multiplicity 12
Kojeve, Alexandre 184
Kolbert, Elizabeth 5
Krementz, Jill 181
Kricher, John C. 32
Krieg, C. Parker 11, 97
Kundera, Milan 1–2

land: as ground of ethics and politics 61; as original literature 26–7; reading 27–8
landscape 26–8, 105, 110, 135, 138–9
"Language of Paradox, The" (Brooks) 103
lastness 52–3
Latour, Bruno 10, 83, 86–7n1, 102, 131, 160, 177
left-wing reader 179, 181, 184
Le Guin, Ursula K. 20
Leitch, Vincent B. 22
Lentricchia, Frank 3–4
Leowald, George 117
Lesjak, Carolyn 49
Lewis, Simon 3
life, as worthy of being lived 182
lightness 1–3
Linebaugh, Peter 124
literary scholarship 26, 77, 176
literary studies 17–18, 77–8, 98n17, 176; colonialist practices mirrored in 26; crisis in 77; and cultural criticism 133–4; relevance, discussions of 17, 48; resistance to instrumentalism 41
literature: fastens language to life 18–19; land as the original 26–7; political contexts of 175; as practice of being in relation 22–9; text itself as 23; value of 4–5, 18–19
"Literature and Ecology: An Experiment in Ecocriticism" (Rueckert) 31
lived experience of reader 83–4
Long, Heather 85–6, 90n20
Long, Mark 8, 9
longue durée 164

looped processes, seeing and saying as 6, 16
Lopez, Barry 26, 27–8
Lorentzen, Christian 20
loss 9, 18, 78, 110, 113, 152, 184–5; lightness of perpetual 1–2; and memory 169; song, as elegaic 15
Lost Words, The (Macfarlane and Morris) 18–19, 27
Love, Heather 89n17, 132, 141

Macfarlane, Robert 10, 18–19, 27, 134, 141n1
Magna Carta 124–5
"Man": Anthropocene made in image of 9, 60; collapse of multispecies agency into concept of 7–8, 60; crisis of 177; as own ghost 8, 62; as unreliable narrator 60–1, 66
managerial ethos 182
Manhattan 63, 64
manifest image 167
Man Without a Country (Vonnegut) 180
Marcus, Sharon 88n14, 146
Marini, Amelia 6–7, 9
Marr, Timothy 64
Marxism 176
Marxist esthetics 34–5
Marx, Karl 62
Marx, Leo 67
masculinist viewpoints 47, 137–8
Masefield, John 11, 159, 162–74; "A Creed" 168–9; reincarnation, interest in 167–9; *see also* "Passing Strange, The" (Masefield)
Maslin, Mark 3
material criticism 77
materiality: of artwork 129; of inhuman 62, 70, 93–4; of "universals" 104
mathematic models of language 145–6
matsutake mushroom 148, 150
McCall, Dan 63, 68, 72n34
McCaw, Neil 8, 94
McFarland, Sarah E. 6
McKibben, Bill 18
meaning: extraliterary 7; false binary relationships 88n14; and form 36–9; and history of reception 82–3; and nostalgia 2; and substance 68–70; unified 22
meaning-making 20–1, 70, 80–1, 85, 102, 108, 176; Bartleby as ground of 63; close reading as 81, 83, 88n13
measurement 7–8; politics of 11, 63, 67, 70
Melville, Herman: balance in works of 61, 67–9, 72n33; "The Encantadas, or Enchanted Isles" 60, 64, 71n6; "Hawthorne and His Mosses" 67, 72n32; *Moby-Dick* 67–8, 72n31, 107; oceanic imagination of 64; *see also* "Bartleby, the Scrivener: A Story of Wall Street" (Melville)
memory 9, 95–6, 169
Menely, Tobias 77, 79, 87n3
Merola, Nicole 7, 78, 87n6
metabolic activity, consumptive nature of 53–4, 54n2; cannibalistic power 64
metanarrative 39
metaphysics, substance as thesis of 69
method 77, 87n3, 132, 144
Michelangelo 122
Middleton, Peter 76
Millar, Liz *152*, 152–3
millennium 1
Milton, John 31
mimetic faculty 156–7
mining histories 140
Mitchell, David 168, 173n9
Moby-Dick (Melville) 67–8, 72n31, 107
modernism 34–5, 41, 103, 116–18; and pacifism 116
modernity 2, 62, 93, 140, 177–8
Moi, Toril 133, 137
"Moon Jelly" (Pagh) 7, 39–40
Moore, Marianne 103
moral accounting 167–9
Moretti, Franco 4, 146, 151, 173n5
Morris, Jackie 18–19, 27
Morton, Timothy 138
mosses 148–50
motivated form, myth of 7, 31; defamiliarization 34, 36; myth of motivated ecopoetic form 31–4; realist and modernist alternatives 34–5; superstructure and base of society 35; and Whitman's poetry 33–5
muddling, concept of 171
Murdock, Iris 112
Museum of Modern Art (MoMA) 117
Museum Wormianum (Worm) 52
Mushroom at the End of the World, The: On the Possibility of Life in Capitalist Ruins (Tsing) 148
mushroom cloud 119, 153–4
mushrooms 148, 150
mystery 19, 28, 66, 137, 140–1

Nadzam, Bonnie 102
Nakashima, George 9, 116–30; Altars for Peace 116, 120–30, *121*, *125*, *126*; butterfly key and imagery 120, 121,

121; concentration camp experience 119; dissent, acts of 129; free edges technique 116–17, 120–1; "One Man's Answer" 118–19; *The Soul of a Tree* 118–19; vision for Altars of Peace 123–4
Nakashima, Mira 128
naming 19, 102, 107; agency of 96–7
narrative: agency of 95; evolutionary 11, 175, 183; grand 177; nonhuman 77, 182–3
narrative archeology 94–100
narrator: modernist versus realist 34–5; nonhuman 182–3; unreliable 60–1, 63
National Geographic 110, 121, 125–6
natural history illustration 47, 50, 52
nature: close reading of 104–7, 151; colonialism imbricated in 94; Hobbesian state of 123; human management of 102; as term 141n1; as unmanageable 102
Nature (Emerson) 18
nature/culture binary 89–90n19
nature writing 136, 138, 141n1, 145, 148; spiritually-inflected 150–1
"negentropic" qualities attributed to poems 31–2
neoliberalism 61–2, 183; and university 176–9
networks 4, 12; actor-network theory 86, 89n17; mycorrhizal 25; trade 53
New Age spiritual travelogue 140
New Criticism 21–2, 26, 76, 82, 87n7, 89n16; attention, focus on 103–4; mathematic models of language 145–6; as "poetics of Fordism" 176–7; *see also* close reading
New Hope *Gazette* 123, 125–6
new materialist theory 135
new nature writing 141n1
New Poetics of Climate Change, The (Griffiths) 40–1
New Theory for American Poetry: Democracy, the Environment, and the Future of Imagination (Fletcher) 32–3
Nietzsche, Friedrich 2
Nigeria 9, 95–100
Nigerian Regiment 96
nihilism 172
Nihil Unbound (Brassier) 172
Nixon, Rob 77
Nobel Committee 127
nonfiction works 10; extinction-focused 46
northern gannet (*Morus bassanus*) 50, 51
nostalgia, aura of 1–2
novel 89n19; changing form of 7, 34; as narratives of identity 89–90n19; scale of 161–2; *see also Burma Boy* (Bandele); *Galápagos* (Vonnegut)
nuclear age 9–10, 119–21, 153–4; arsenals, failure to maintain 128; mutually assured destruction 122; semiotic event of bomb 120; and urban planning 119–20; Vonnegut on arms race 182

object 36; as anchor 110–11; close reading of 4; literary 79, 81, 83, 104, 176; nonhuman reduced to 26
objectification 18, 26, 29
Occupy movement 63, 69
Olaoluwa, Senayon 8–9, 10
Olkiluoto Island (Finland) 136
"One Man's Answer" (Nakashima) 118–19
On the Morning of Christ's Nativity (Milton) 31
Opperman, Serpil 77, 78
originary technicity 178–9
origins 138–9
Orr, David 41–2
orthodoxy, challenging 21
Ostergard, Derek E. 117
ostranenie (defamiliarization) 34, 36, 42
Oxford Junior Dictionary 18–19

Pagh, Nancy 7, 39–40
panic syndrome 61
Parham, John 5
"Passing Strange, The" (Masefield) 11, 159, 162–74; abstracted nature of verse 163–4; accelerated time scale 159, 164–5; anti-mystification, tone of 163; earth processes in 169–71; grand scale and moral accounting in 167–9; opening lines 162, 165
Paterson, Don 37–8
patterns 145–6, 156–7
pedagogy 5, 20–4; eco- 40–3; ends-based thinking 28; labor of devalued 80; reading as transferable skill 82; reverse engineering 37; staged education 27–8; STEM-centered eco-literacy 41–2
personhood 167–8
petrol discourse 98–9
Phaedrus (Plato) 178–9
phenomenon-focused approaches 78
Phillips, Dana 32
philology, return to 177–8
phrase 33
Pilgrim at Tinker Creek (Dillard) 150

planning 181, 184
plant studies 9
Plato 178–9
Pleistocene 104
poetics: Anthropocene 76, 102; eco- 7, 31–4, 40–3, 114n3
poetry: Cambridge study of response to 30–1; meter 38–9; "negentropic" 31–2; poem as environment 33, 35; scale represented in 161–2, 164–5; *see also* Bishop, Elizabeth; ecopoetics; Masefield, John; Stevens, Wallace
Poirier, Richard 24
political polarization 17
politics, responsibility as 70
"Position of the Narrator in the Contemporary Novel, The" (Adorno) 34
postcolonial remains 8, 94–5, 97
postcolonial studies 8–9, 94
posthumanism 77, 175, 177–8
postmodernism 95, 171, 177
powerlessness, valorization of 184
Practical Criticism (Richards) 30
Pragmatism (James) 19
pragmatist close reading 6, 16, 22–4
precision crafting, rhetoric of 37
Prize Designs for Modern Furniture competition 117
professional class intellectuals, complicity of in planetary crises 175, 182
proletarianization of thought 179, 186
proximity 8, 52, 78, 80, 132, 161

radical epistemologies 10, 134–5, 140, 141
Randall, John H. 68
Ransom, John Crowe 4
rationalism 22–3
Ray, Sarah Jacquette 80
reader-response theories 82
reading: across disciplines 48–9; across natural and cultural histories 77–8; conjunctive 11, 145, 151–7; as *in medias res* 82; justifications for 17, 79; lay and professional 78; literary experiences versus passive consumption 20, 88n10; nonhuman 183–4; as noun 184; as provisional and iterative 49; and scale 160–6; solitary reader, trope of 85–6; stratified 139–40; surface 49, 81, 88n14, 132–3, 137, 146–7; as transferable skill 82; witnessing 55n14; *see also* close reading
Reading: Acts of Close Reading in Literary Theory (Wolfreys) 165
reading groups 90n20
Reagan, Ronald 127
realism 160–1, 167
reality 69–70, 72n36, 89–90n18; in Bishop's poetry 108–9, 113; and narrative archaeology 95, 96
reception history 82–3
reference 34–5
reincarnation 167–9
relationship 69–70; close reading as relational activity 8, 19–20, 22–9; environments as natural ensemble 33; with nonhuman world 25; surface reading, critique of 49
representation 11, 34, 159–61, 165; cultural 61, 78, 134
representatives, political 33
responsibility 2, 18, 39, 61–2, 69–70, 96
"Rhetoric of Climate Change" course 16–17
Richards, I. A. 7, 30–1, 37
Rilke, Rainer Maria 162
Robinson Crusoe (Defoe) 110, 184
Romantic writing 32, 47, 103, 105; Stevens's critique of 108; temporal dissonance in 76
Rosenblatt, Louise 81–2
Ross, Andrew 181
Royal West African Frontier Force 96–7
Royal West African Frontier Front 94
Ruddiman, William 3, 13n6
Rueckert, William 31, 114n3
Russian Formalism 35–6

Sachs, Jonathan 76
Said, Edward 96–7, 177–8
Salk, Jonas 135
Sarewitz, Daniel 171
scale 4, 5, 7–8, 87n3, 87n5; antagonism overdetermines projections of 175; derangements of 11, 79, 179; derangements of in "The Passing Strange" 159–74; of embodiment 160; as fundamental structure 159–60; grand 167–9; immensity of world 28–9; in individualist rhetoric of liberalism 87–8n9; long-term, applied to ecosystems 32; *longue durée* 164; micro- and macroscales 159; move to antagonism from 175, 179, 186; "normal" 160–1, 167; perception of 159–60; and personal experience 18; poetic resources for representing 161–2, 164–5; and reading 160–6; reconnecting human life with 5; of

representation 134; scalar gap 61–2; spatial 98, 138, 159, 162, 165–7, 179; of time 8, 76, 78
scale critique 11, 159, 161
scale effects 132, 179
scansion 41
Scarlet Letter, The (Hawthorne) 23
Scarry, Elaine 124
scenario planning 48, 184
Schell, Jonathan 122–3, 127
Schuster, Joshua 47
science fiction 160, 177, 185
sciences 4, 18, 20–1, 25–6, 148, 178
scientism 178
Scroggins, Mark 105, 114n4
Searle, Leroy 82–3
Sebald, W. G. 138, 141–2n2
Sedgwick, Eve Kosofsky 4, 26
seeing, learning new ways of 149–52
Selby, Nick 114n3
Sellars, Wilfrid 167
semiotic metaphors 48
sensory awakenings 16
"sensuous apprehension" 30, 43
sentence 23, 33
Shakers 121
Sharma, Meara 19
Shklovksy, Viktor 34, 35–6
"signature poems" 36
signification 2, 22, 65
similarity 156–7
Singularity of Literature, The (Attridge) 41, 42
situatedness 84
Six Memos for the Next Millennium (Calvino) 1, 12
Sixth Mass Extinction 4, 13n8, 60
Skinner, Jonathan 42
slavery *see* Atlantic Slavery
slow reading practices 4, 39, 42
slow violence 77
Smith, Barbara Hernstein 81
Smith, John 7, 37–9
Smith, Zadie 24–5
Snyder, Gary 2, 4, 62; "Language Goes Two Ways" 40
social contract 65, 116, 120, 123–4, 128–9
social difference, elided under *Anthropos* 11, 175
Socialist Realism 34
solar eclipse *see* total solar eclipse
song, as elegaic 15
Soul of a Tree, The (Nakashima) 118–19
Soviet Academy of Art 128
Soviet Union 34, 128
spatial metaphors 81, 159
spatial scales 98, 138, 159, 162, 165–7, 179
species thinking 94, 96, 140
specimen collection 46
speculation 48, 62, 70, 148, 172; in *Galápagos* 177, 180, 183–5
Sri Aurobindo 128
Staying with the Trouble: Making Kin in the Chthulucene (Haraway) 15
STEM-centered eco-literacy 41–2
Stevens, Wallace 9, 102–15; analogy in writing of 107–8, 110; "Anecdote of the Jar" 171–2; drive to cut through "layers of varnish" 106, 108; "Evening without Angels" 113; "Fabliau of Florida" 108–9; "Farewell to Florida" 104; "The Idea of Order of Key West" 109–10; "Indian River" 104–5, 114n4; "Infanta Marina" 109; "Nomad Exquisite" 106; "O Florida, Venereal Soil" 106
Stiegler, Bernard 178–9, 186
Stirner, Simone 146–8, 156
Stoermer, Eugene F. 1
stratigraphic signals 2–3, 48
Street, Laura-Gray 37
subject: close reading of 4; grievable 52, 54
subjectivity 167–8; black, deformation of 94; of plants 26
sublime 11; and eclipse narratives 149–50, 156; and *Galápagos* 175, 178, 183; Melville's oceanic imagination 64
substance 68–70
"Sun and the Fish, The" (Woolf) 10, 151–2, 154–7
surface reading 49, 81, 88n14, 132–3, 137, 146–7
symptomatic reading 10, 11, 58, 88n14, 132–3
system of systems 12
systems-thinking 31

Taylor, Jesse Oak 77, 79, 87n3
technicity, originary 178–9
temporal dissonance 76
Tennyson, Alfred 162
terror: affect of 67–8; visual imagery of 119
text: allowing to be close 23–4; attention to 22–3, 133; critical work done by 104; embedded in the nonliterary 83; engagement with as text itself 21–4, 137; singular focus on 82; unseeing 145–6, 148
theory-led and identity-centered criticism 41
Thomas, Aaron 46, 50, 53

Thompson, Hilary 10
Thoreau, Henry David 22, 69
thought: balance as problem of 67–9; proletarianization of 179, 186; relationship to world 69
Thus Spoke the Plant (Gagliano) 18, 26
time: accelerated in "The Passing Strange" 159, 164–5; deep 51, 64, 78, 86n1, 135–9; disjunctions in 165; flow 27; human acceleration of geological change 61, 71n6; out of joint in Anthropocene 48; paradoxical temporality 76; scales of 8, 76, 78; temporal dissonance 76; time-space compression 132
"Tithonus" (Tennyson) 162
"Total Eclipse" (Dillard) 10, 151, 153–4
"Totality: The Colour of Eclipse" (Carson) 10, 154–6
total solar eclipse 10, 144–58, *152*; eclipse writing tradition 146–8, 151–3; human-centered writing about 153–4; non-human-centered writing about 154–5; reading practice in constellation with essays about 148
transactional approach to reading 82–4, 86
transatlantic exploration and trade networks 53
transcendence 62, 69, 109
transcendentalism 64
translocation of species 13n8
travel genre 136–8, 141n1; New Age 140
"traveling theory" 96–7
Travisano, Thomas 107, 112, 114n7
tree communication 25
trees 9–10; as archive 116–30; close reading of 119–21
Tsing, Anna Lowenhaupt 10, 148, 149–50
twentieth century, middle of 2–3

Ulloa, Astrid 25
unbearability of world 1, 3, 5–6, 12
Unbearable Lightness of Being, The (Kundera) 1
uncanny 10, 40, 137, 141
undecideability 177
Underground (Hunt) 10, 134, 138–41
Underland (Macfarlane) 10, 134, 135–8
UN Intergovernmental Science-Policy Platform on Biodiversity and Ecosystem Services 5
universalism 4, 8; attributed to "innate biological rhythms" 139; of left thinkers 178; materiality of 104; of species thinking 94, 96
university, neoliberal 176–9; ideological containment in 179
unknowing 177
unreadability 179
unseeing 145–6, 148–50, 156
unthinkability 5, 70, 85, 90n19, 182
urban inhumanity 64
urban planning 119–20

Vermeulen, Pieter 159, 161, 182–4
Vertebrate Fauna of the Outer Hebrides, A (Harvie-Brown and Buckley) 51
verticality 135
Vietnam War 121, 122, 180, 183, 186
violence 177–8; cross-species conflict 47; forms of 46, 54n2; of great auk extinction 46–7, 50, 52–4; hunting, narratives of 52; racialized 182; slow 77
visual imagery, of terror 119
Vonnegut, Kurt 11, 175–89; "Fates Worse than Death" 182, 185; "freshwater socialism" of 180–2; "Preface to Galápagos" 181

Wajarri people (Western Australia) 140
Walden (Thoreau) 69
Wald, George 9, 122–3, 126, 127–8
"Wallace Stevens and Elizabeth Bishop: Ideas of Order" (Gelpi) 102–3
walnut tree 123–4
war 2, 11, 186–7; Cold War concerns 153–4; Nakashima's Peace Altars as response to 119–24
"War Against the Center" (Galison) 119–20
Watson, Robert 39
Waugh, Charles 64
"Way We Read, The" (Guillory) 78
We Gotta Get Out of This Place (Grossberg) 176
Weimar constitution 120
Whitbourne, Richard 45, 46
"White Geology" (Yusoff) 8, 93–5
White, Stewart Edward 150
Whitman, Walt 33–4, 35
wild systems, languages as 40
Williams, William Carlos 36
Wimsatt, W. K. 145–6
"Witch of St. Kilda, The" (Ford) 47, 48, 51–4
witnessing 55n14
Wolfreys, Julian 165
Wood, David 164, 167, 169
Woods, Derek 159, 161
woodwork 9–10
Woolf, Virginia 10, 151–2, 154–7

words: inhabiting 6–7, 16, 21, 29; at sentence level 23; *see also* inhabitation
"Work of Art in the Age of Its Technological Reproducibility, The" (Benjamin) 145
world: administered 34; age of the world target 9, 116, 119, 122, 128–9; immensity of 28–9; relationality with nonhuman 25; relationship of thought to 69; unbearability of 3, 5–6, 12; *see also* Earth
World Beyond Your Head, The (Crawford) 84
"world risk society" 137
World War II 34, 94, 96; deceptive recruitment 97–8; impossibility of closure to memory of 9, 100
Worm, Ole 52
Wright, Christopher 93, 96
writer, trope of 85
writing 55n11, 55n14, 85; as close reading 5; nonanthropogenic forms of 78; as provisional 49

Young, Richard 8
Young, Robert 94–5
Yu, Beongcheon 67
Yusoff, Kathryn 2, 10, 62, 70, 95, 139–40; "White Geology" concept 8, 93–5

zeitgeist 177, 184
Zhukov, Yuri 127
Žižek, Slavoj 175
"zooming out," as transcendence 62

For Product Safety Concerns and Information please contact our EU
representative GPSR@taylorandfrancis.com
Taylor & Francis Verlag GmbH, Kaufingerstraße 24, 80331 München, Germany

www.ingramcontent.com/pod-product-compliance
Ingram Content Group UK Ltd.
Pitfield, Milton Keynes, MK11 3LW, UK
UKHW020006200926
475702UK00015B/2498
9780367466596